CELL PHYSIOLOGY AND PHARMACOLOGY

CELL PHYSIOLOGY AND PHARMACOLOGY

by

JAMES FREDERIC DANIELLI

Ph. D. (LONDON AND CAMBRIDGE), D. Sc.

*Professor of Zoology, King's College, London,
and Honorary Lecturer in Pharmacology, University College, London.
Formerly Reader in Cell Physiology, Royal Cancer Hospital, London,
and Fellow of St. John's College, Cambridge*

(Facsimile of the 1950 edition)

HAFNER PUBLISHING COMPANY
New York and London

1 9 6 8

Originally published 1950
Reprint 1968

Printed and published by
HAFNER PUBLISHING COMPANY, INC.
31 East 10th Street
New York, N. Y. 10003

14806
Library of Congress Catalogue Card Number: 68-21495

Printed in U.S.A.

PREFACE

This book is based upon lectures given at University College, London, at the invitation of Professor F.R. WINTON. It was not intended that these lectures should constitute an exhaustive survey of the place of cell physiology in pharmacology. They were intended to indicate some of the more important factors, on the cellular level, which must be considered by students of drug action.

My interest in the mode of action of drugs has been stimulated at various times by A.J. CLARK, H.R. ING, J.C. DRUMMOND, J.H. GADDUM, and F.R. WINTON. During the war of 1939–45 this interest was turned towards practical problems, and I was able to study at first hand the problems which are encountered when a search is being made for a new drug. The problems roughly separate into three broad groups—corresponding to the disciplines of organic chemistry, physical chemistry, and biology. It was impressive to see that the chemical problems could often be handled with ease and moderate precision. But in practically all investigations it becomes necessary for the chemist to introduce theorisation on the biological side at some stage. This is sometimes done as a guide to action, and sometimes as a guide to inaction. But whatever the intention, the chemist's biological theories are apt to be more surprising than successful.

When, as sometimes occurs, the biologist is called in to deal with the biological sides of the problem, his success is frequently limited to pointing out errors in the chemist's biological theories. The reason for this limitation is the paucity of information on the biological side. Whereas biology is probably more complicated than chemistry, vastly more man-hours have been devoted to chemical research than to biological research. When this situation is remedied, it will be possible for the biological side of drug development to proceed in a rational manner. This will eliminate the hit or miss technique, which is now the main basis of the search for new drugs. For chemists this will mean much diminution of effort—probably 90% or more of the time used for synthesis of useless compounds can be eliminated. And for biologists there will be a corresponding diminution in the time spent in testing inactive compounds.

Consequently my hope is that these lectures will play some little part in directing attention to the biological aspects of drug action, and in showing that studies in the biological field are practicable.

I am deeply indebted to Prof. A. HADDOW for advice and for reading the galley proofs.

J. F. DANIELLI

CONTENTS

Preface v

I. The Cell as a Physico-Chemical Unit

Introduction 1
Cytochemical aspects 2
Cytoplasmic gels 4
Chromosome and gene structure 7
Physico-chemical aspects 11
Defence mechanisms 20
Self-reproducing bodies 22
Integration 23
References 24

II. Possible Actions of Drugs on Surfaces

Ionic interactions 25
Dipole interactions and complex formation 32
Specificity in surface reactions 34
Mechanism of lysis 36
The action of oestrogens on monolayers 37
The effect of micelle formation 39
Long-range forces 43
References 45

III. Membrane Permeability and Drug Action

Introduction 46
Membrane permeability and drug structure 48
Problems of the access of drugs to organs 63
Examples of the permeability factor in drug action 66
References 73

IV. Enzymes and Drug Action

Functions of enzymes 74
Possible functions of drugs in relation to enzymes 75
Problems in the analysis of the action of drugs on enzymes . 78
The action of drugs on respiration and glycolysis in muscle . 82
The action of various enzyme poisons on different physiologi-
cal processes 85
Classification of drugs according to their physiological effect 87

The classification of drugs in terms of enzyme systems upon
which they act 89
The mode of action of vesicants. 91
Biological aspects of enzyme studies 94
References 96

V. The Actions of Narcotics

Introduction 97
Theories of actions upon surfaces 99
Theories based on oil-water partition effects 105
Theories based on actions on enzymes 109
References 114

VI. Responses of Cells on the Biological Level

Introduction 119
The nature of biological responses. 119
Artificial parthenogenesis 122
Mitotic poisons 123
Reproduction of bacteria and viruses 136
Nuclear and cytoplasmic drug action 139
Possible modes of drug action upon genes 141
The relationship between hormones and evocators 146
References 149

Author Index 151

Subject Index 153

above are considered, it becomes at once clear that the physico-chemical structure of the cell is in a dynamic condition. The fact that the broad outlines of structure, as shown by classical histological methods, are not readily changed, is indicative of certain stable elements amongst those determining cell behaviour. But this should not hide from us the great variation which may occur on the chemical level. This variation occurs at just that level on which we must expect most drugs to act, so that in a sense the variability of the cell at this chemical level is of more importance to us than the relatively static nature of certain structural patterns.

It is possible to procedure redistributions of most substances in the cell without irreversibly damaging the cell. E.g. when a cell goes into division, the organisation of some of the cell products is totally changed in the production of a spindle. The sharp distinction between nuclear sap and cytoplasmic material breaks down. Alternatively, by centrifuging the cell, we can obtain a new distribution of substances quite unlike that which arises in a cell under normal conditions or when it is dividing. If the use of centrifugal force is not excessive, the cell can reconstitute the original organisation of substances, and continue with its normal functions unimpaired. In the case of *Ascaris* eggs, the unfertilised eggs may be centrifuged at 100,000 times gravity for four days, and yet still develop normally when fertilised. Thus particular adjuxtapositions of matter which are shown up by cytochemical investigations are not all of them physically necessary for the maintenance of the life of the cell. Insofar as there are absolutely unchangeable centres of organisation in the cell, these must be based upon the formed bodies of the cell such as chromosomes, mitochondria, the cell and the nuclear membranes, and the cell granules. The evidence at present available shows that as long as these bodies are left intact, the cell can recover from quite violent treatment and disturbances of its normal chemical condition. But this does not mean that the distribution of chemical substances which we see in the cell under normal conditions is of no consequence. Although cell constituents may be rearranged by centrifugation or by micro-dissection procedures without killing the cell, there is a certain amount of clear evidence showing that the activity of the cell is no longer carried on in the same way during the period in which this redistribution exists. Thus BRACHET has shown that the cyanide-sensitive respiration of amphibian liver is eliminated by stratification in the centrifuge, but returns when the cell constituents have been restored to their normal positions.

Cytoplasmic Gels

Except during mitosis, protoplasmic gels do not play a very obviously prominent part in the activity of hepatic cells. But in a variety of other cells experimental procedures have shown that cytoplasmic gels are responsible for many characteristic activities. The normal condition of these gels is essential for mitosis and cell division, protoplasmic streaming, amoeboid movement, phagocytosis and the maintenance of cell form. For example the marine protozoon *Ephelota coronata* throws out fine protoplasmic tentacles, the length of which is many times greater than their diameter. This structure is maintained by the cortical gel layer which lies just inside the plasma membrane in the tentacles. When this gel is liquified by application of high pressure, the operation of surface tension forces immediately breaks the tentacles up into a series of fluid droplets. The importance of organised structures is also shown by some experiments which were conducted by MOORE on the slime moulds. These moulds were able to grow through membranes having a pore diameter of the order of 1 micron. But if they are filtered by pressure through pores as large as 100 microns in diameter, the respiration of the mould is decreased by 50%, and if the pore size is not larger than 20 microns, the organism is killed. In considering the significance of these gels in the daily life of the cell, it is important to remember that cell division, protoplasmic streaming, amoeboid movement, phagocytosis, and the maintenance of cell form can go on quite normally in the absence of the nucleus. In some cases the absence of a nucleus does not exert a pronounced effect for several days after its removal from the cell.

There is comparatively little information about the details of structure, and so forth, of these gels in most cells. In striated muscle cells, the gels become so prominent as to be the characteristic feature of the cell, and it is known that the component molecules are arranged in organised fashion to produce definite intracellular bands, discs, fibres and membranes. The most striking of these features of the living cell is the distinction be-

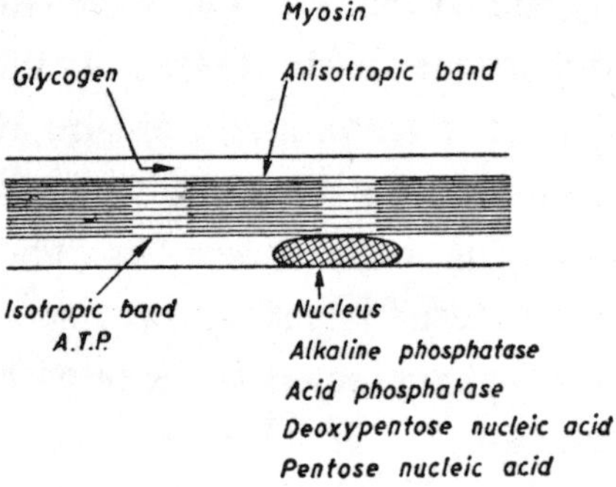

Fig. 2. Diagram of distribution of substances in a striated muscle fibre

tween anisotropic bands, which presumably contain the greater part of the myosin of the cells, and the isotropic bands. The former bands rotate the plane of polarisation of plane polarised light, whereas the isotropic bands do not. Cytochemical studies on muscle cells are at present rather scanty. It is known that adenosine triphosphate is concentrated in the isotropic bands, and that these bands also contain a relatively labile alkaline phosphatase. Whereas acetone resistant acid phosphatase and desoxyribose nucleic acid are both mainly present in the nucleus. This is indicated diagramatically on fig. 2. We may be quite sure that this localisation of chemical substances has special chemical and physical significance. One thing which is very clear is that often this localisation keeps substrates apart from the enzymes which would otherwise destroy them or utilise them prior to the onset of the physiological activity for which this process is useful. For example the co-existence of phosphate esters and of phosphatases which are active at cellular p_H is only made possible by this physical separation of the substances concerned. As soon as the biochemist, by grinding or other maceration procedures, breaks down this organisation on the chemical level, phosphate esters are rapidly destroyed, and with them the characteristic cycles which are involved in glucose metabolism.

Chromosome and Gene Structure

The structures about which most details are available are the chromosomes. There are two very different main theories of the structure of the chromosome. One is that it consists of at least one polypeptide chain running the full length of the chromosome, with discrete bodies, known as chromomeres, distributed along it. The chromomere, according to this theory, consists of material additional to that contained in the long polypeptide chains. Each chromomere has at least one gene and may contain several genes. According to the other theory, the normal chromosome contains one or more polypeptide chains extending the whole length of the chromosome. The chromomeres, however, are believed to be not additional material, but regions in which the main polypeptide chain is much folded upon itself. As in the first theory, each chromomere is believed to consist of one or more genes.

As STEDMAN & STEDMAN and MIRSKY have shown, chromosomes consist of at least two types of protein, the basic proteins known as histones and acidic proteins which STEDMAN calls chromosomin, and also two types of nucleic acid, one of which contains a pentose and the other a desoxypentose. The information about the physical structure of chromosomes has been very largely derived from studies of exceptionally large chromosomes, such as those found in certain plant cells, and the giant salivary chromosomes of the *Diptera*. Whichever of the theories of chromosome structure may be correct, it is certain that the total amount of matter in the chromomeres is greater than that in the parts of the chromosomes lying between the chromomeres. Cytochemical studies have shown that by far the greatest part of the purine and pyrimidine, the desoxy sugar, tyrosine, histidine, tryptophane, and alkaline phosphatase are present in the chromomeres. The interbands are relatively lacking in these substances. Since genes are known to be located in the chromomeres, it seems pro-

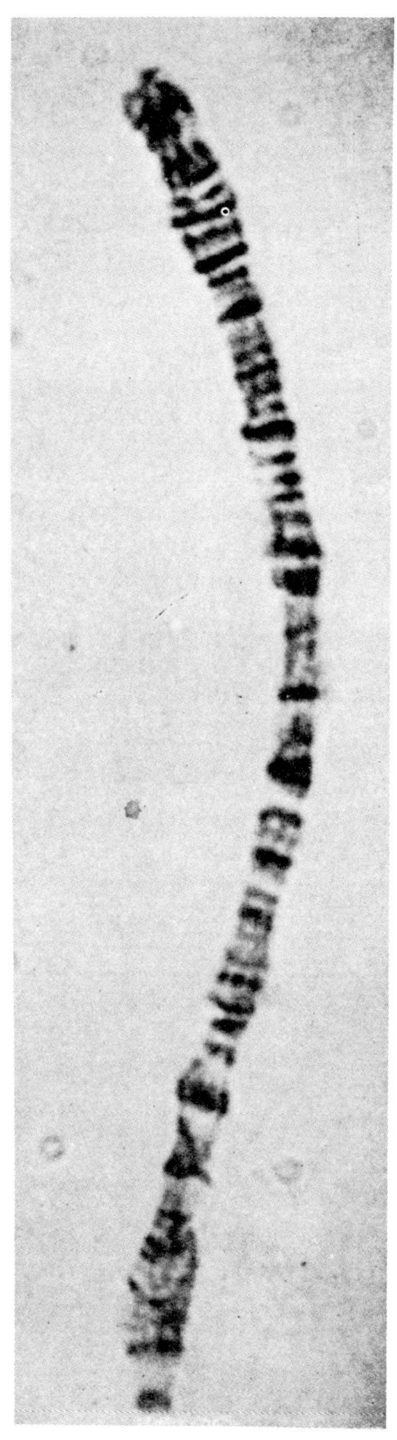

Plate I. Distribution of alkaline phosphatase on an X-chromosome of *Drosophila melanogaster* (DANIELLI and CATCHESIDE).

bable that the substances which have just been mentioned are essential constituents of genes: in fact it is likely that a gene can be considered as a special array of enzyme molecules, organised so as to produce one chemical product, or a small family of chemical products. The substances so produced are those involved in the mediation of the effect of the gene. In addition to these substances, the gene has the capacity to reproduce itself completely. The extent to which this is a different function from that of producing the substance or substances concerned in the mediation of the genetic effect is unknown.

The sharp localisation of chemical substances in the chromomeres, and the variation in concentration of these different substances from chromomere to chromomere, must involve also, through the operation of Donnan equilibria etc., highly local variations in p_H along the chromosome and probably also in the SH content of different parts of the chromosome. These two physico-chemical factors must be very important to consider in relation to the enzymic activity of a gene.

Physico-Chemical Aspects

The foregoing discussion has been based mainly on consideration of the cytochemical distribution of different substances. In forming a picture of the physico-chemical system characteristic of the cell, we must consider a number of other properties of the molecular types constituting the cell. The most important points to be considered here are: 1. the units of structure, 2. the control of enzyme systems, 3. the dielectric properties, 4. membrane properties, 5. the dynamic condition of cell constituents.

Units of Structure. Fig. 3 shows diagrams of a number of types of units of structure which will be formed by

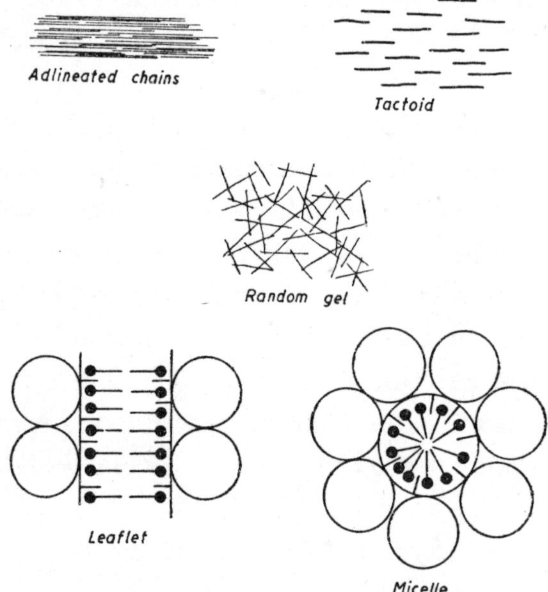

Fig. 3. Units of cellular structure

molecules known to be present in the cell. The simplest of these consists simply of adlineated polypeptide chains such as are found in collagen fibres, keratin, muscle fibres and chromosomes etc. In the cases of the chromosome fibre, muscle fibre and collagen fibre it is known that there is a good deal of differentiation along the length of the chains. It is likely that this is also the case with most other natural fibres. Except in the case of the chromosome, it is not at all clear what significance should be attached to the local differentiation along the length of the fibre. One theory is that it assists in the precise adlineation of polypeptide chains, thereby giving rise to fibres of maximum strength.

A physically quite different type of structure, also composed of adlineated protein molecules, is the type found in tobacco mosaic virus tactoids. These consist of needle-shaped molecules, which are oriented parallel to one another, but which are not in contact at any point. They are maintained in these positions, with the distance between the long axes of the molecules relatively well defined, by long range forces whose nature is not yet clearly understood. BERNAL has supposed that the spindle and asters of a dividing cell are composed of such tactoids. It is also possible that the forces acting between daughter chromosomes during mitosis are of this type, that cell adhesions and the form of cells are in part determined by such forces, and even that the differentiation of cells is influenced by forces of this type.

In addition to the gels or tactoids mentioned above, there are the cortical gels which do not seem to consist of very highly oriented molecules. They may have a closer resemblance to gelatin gels in which the individual molecules appear to be dispersed more or less at random.

When the lipoid molecules of the cell are also considered, we have units of structure of a fresh type arising, based on the micelle and on bimolecular leaflets of fatty molecules. These types of structure are also shown in Fig. 2. The stability of these structures is based on the fact that the hydrocarbon parts of the molecules concerned are in effect squeezed out of solution because water attracts water more strongly than it attracts hydrocarbon. The polar groups of the molecules become anchored in the interfaces between the hydrocarbon parts of the molecules and the water. Owing to the very high surface activity of protein molecules, the surfaces of lipoid micelles and bimolecular leaflets must, under biological conditions, always have adsorbed upon them a monolayer of denatured protein. Under most circumstances there will also be adsorbed upon this primary layer of denatured protein a secondary layer of globular protein molecules. Thus a very complex structure may arise simply as a result of the operation of adsorption forces.

So far, the structure of the cytoplasmic granules found in the cell has not been made sufficiently clear for detailed discussion. Nor is it clear what part the different nucleic acids play in determining the structure of the chromosomes and other bodies in which they are found.

It is very important to notice that as a result of the presence of granules, chromosomes, membranes, nucleoli and other formed bodies in the cell, there must necessarily be great importance attributed to surface properties. The main molecular constituents of cells, the proteins and nucleic acids, are themselves such large molecules that any reaction taking place with them, or in which they are involved, is necessarily a surface action and not a bulk reaction. And it must very often be the case that two reacting molecules must be regarded as reacting in the zone constituted by their overlapping surfaces. In considering the basic physico-chemical nature of the cell, we must therefore be particularly alive to the importance of surface properties.

The Control of Enzyme Systems. Amongst the most important of the physico-chemical systems which form an integral part of living cells are the systems controlling the activity of intracellular enzymes. Among the factors involved in controlling enzymes are hydrogen ion concentration, the concentration of SH groups, the concentrations of inhibitors and activators, and those processes which control access of substrates to enzymes.

The hydrogen ion concentration is maintained constant in cells partly by the buffering substances normally present, and partly by the active excretion of excess of acidic or basic substances. The SH content of a cell is important because many enzyme systems have maximal activity when either in the reduced state or in the oxidised state. Usually the reduction can be carried out by SH compounds, particularly by glutathione. Similarly, the reduced forms of enzymes can usually be oxidised with the oxidised form of glutathione. Thus the presence within cells of glutathione constitutes a poising system tending to maintain a given degree of reduction of the enzyme systems of the cell, in just the same way as the p_H buffers tend to maintain a given degree of ionisation of the enzyme systems.

Equally important with p_H and SH content are the factors of inhibition, activation and substrate access. But extremely little is known of the variables which control the operation of these factors.

However, the cell does not present a completely uniform environment with respect to p_H and SH content. Thus, when we consider a gene present in the nucleus, one of the most striking properties is the high concentration of ionising groups in the gene. As a result there is a Donnan equilibrium between the nuclear sap and the gene. Thus the p_H in the gene will not be the same as that in the nuclear sap. Even when we are dealing with small bodies, like protein and nucleic acid molecules, we encounter local variations in p_H. This is because there are high concentrations of ionising groups in the surface of proteins and nucleic acids, so that equilibria analogous to the Donnan equilibrium exist between the fluid medium of the cytoplasm and the surfaces of the protein and the other colloidal particles dissolved or suspended in it. The difference in p_H between such surface and bulk phases may amount to 1 p_H unit or more. Thus the interior of a cell, although its average p_H value may be very constant, can in fact present an extremely variable p_H in different sub-microscopic regions. This permits of great variations in the degree of enzyme activity, according to the particular sub-microscopic environment with which individual enzyme molecules are associated.

Very similar effects exist in connection with the control of the reduction of enzyme systems by glutathione. In addition to the "diffusible" SH groups of glutathione, there are other "indiffusible" SH groups permanently fixed to protein molecules. The distribution of glutathione between the surface of an enzyme, and the surrounding bulk phase, is in part determined by the total charge on the protein and the degree of ionisation of the glutathione. Hence there is a fairly complex equilibrium determining the distribution of SH compounds between the surface and bulk phases within a cell. Furthermore, it may be important that whereas surface p_H is almost independant of the surface SH content, as the result of the operation of electrostatic factors the surface SH content is not independant of p_H.

Dielectric Constant of Cellular Systems. A number of workers have pointed out that the presence of large polar molecules, such as proteins, within living cells will markedly raise the dielectric constant of the interior of a cell, by comparison with the dielectric constant of the external media. This increase in dielectric constant (as measured by relatively low frequency alternating current) is thought likely, by some workers, to have important consequences for the cell. But very little as yet is known about these factors.

Membrane Properties. Extremely little is known of the physico-chemical properties of the nuclear membrane, but much is known about the properties of the plasma membrane of the cell. The most striking properties for our present purposes are 1. its selective permeability, 2. its polarised condition, 3. its asymmetry and 4. the "active patches" present within it.

In a later chapter the permeability of the cell membrane will be dealt with in more detail. It is sufficient for the moment to know that whereas diffusion is rapid through the cytoplasm and through the nuclear sap, permeation of the cell membrane can be a very slow process, and it is possible that this is also true of the nuclear membrane.

The polarised condition of certain cells, such as those of nerve and muscle, is at present difficult to understand in relation to the energy supplying systems which maintain this state of polarisation. But the development of further understanding of this is of vital importance for the analysis of drug action, because as a result of this polarisation, the cell membrane becomes a highly labile system capable of giving almost explosive responses to certain drugs such as acetylcholine, or alternatively of losing its labile character in the presence of other drugs.

Possibly connected with the lability of polarised membranes is the presence in many cells of small patches which are selectively permeable to certain substances. It may be that the action of acetylcholine is not an action generalised over the whole surface of the responding cell, but is confined to small active patches. Similarly the stimulating action of light, which is mediated by visual purple, may involve a sensitive system localised in active patches in the cell membrane.

The Dynamic Condition of Cell Constituents. Studies made with isotopes, though still very far from complete, have shown that practically all the atoms present in the apparently stable structures of a cell are being fairly rapidly exchanged with other atoms of the same type. Thus, although many features of cell architecture when examined by histological or cytological methods may appear stable, in fact every part of the cell appears to be in a dynamic condition. Every enzyme molecule, every protein molecule, every nucleic acid molecule, and probably every part of a gene is in a state of constant change on the chemical level. It is very easy to see that drugs may find sites of action of the greatest importance in places where they can interfere with these processes of degradation and rebuilding of the constituent molecules of the cell. It is probable that permanent interference with any one stage in these processes will eventually cause the death of the cell.

Defence Mechanisms

Many of the defence mechanisms of the cell are physico-chemical in nature. They may in some measure be divided into mechanisms of short term importance and of long term importance. Among the short term mechanisms available to the cell for dealing with foreign bodies are concentration in vacuoles or granules, and detoxication processes. A foreign substance which becomes introduced into the cytoplasm of the cell may be practically removed from most of the cell and concentrated either in vacuoles, as is often the case with neutral red, or in granules as is often the case with trypan blue. As a result of these processes, most of the rest of the cell can function unaffected by the foreign substance. Although these processes are only obvious in the case of coloured substances, there is, of course, no particular importance in a substance being coloured. A colourless substance is concentrated in the same ways as are coloured substances. Then, possibly after a substance has been concentrated in the interior of a cell in this way, it may be subjected to detoxication processes. Often these processes involve the addition to the molecule of a polar residue, such as glucuronic acid or sulphuric acid. As a result of this type of procedure, the chemical and physico-chemical character of the foreign substance may be altered so as to make it comparatively innocuous. Thus menthol in the body is largely converted into menthol glucuronide, which is devoid of toxic properties and also has much less surface activity than has menthol itself. Also the increase in polarity greatly reduces the probability that the detoxicated molecules will re-enter the cell once expelled after detoxication.

An alternative method of detoxication is frequently encountered in the form of destructive reactions, in which the foreign molecule is broken down into smaller and less toxic compounds. These processes are usually enzymic. Thus toxic amines are often detoxicated by amine oxidases.

Amongst the longer term processes which may develop on the physico-chemical level, there are included increased efficiency in detoxication of a given foreign substance, the development of "resistance" to the foreign substance, and the development of antibodies.

The development of resistance probably takes many forms. It may consist of an increase in ability to detoxicate or to destroy a foreign body. This type of resistance bears many formal semblances to the development of adaptive enzymes in bacteria and yeasts. There is also some evidence that the development of resistance may involve changes in the permeability of the cell membrane, reducing the ease with which the foreign substance is able to penetrate into the cytoplasm.

The activity of antibodies may take two rather conspicuously different forms. The antibody may be released from the cell and combine with the foreign body or foreign organism, before the latter reaches the cell. Or alternatively the antibody may be present on the outside of the cell membrane, and react with the foreign body, to prevent its penetrating into the cell.

Two rather obvious conclusions follow from some of the arguments advanced above. The first is that, as the result of secretory activity of the cell, to give the average concentration of a drug in a cell is practically meaningless, since it is very improbable that any drug is uniformly distributed through the cell as a whole. Secondly, there is a competition between the build-up of a substance inside a cell, and the rate at which it is removed by detoxication or other processes. Both these points must receive proper consideration if the action of drugs is to be fully understood.

Self-reproducing Bodies

One of the most characteristic activities of biological systems is their ability to reproduce themselves. The cellular bodies which we at present anticipate may be able to do this, or know are able to do this, are nuclear genes, plasma genes, viruses, and adaptive and other enzymes. A drug may act upon these bodies, preventing their self-reproduction, or causing them to reproduce in a new manner. Either of these activities would be expected to produce pronounced changes in the cell. The problems involved in instances such as these, where drugs are interfering with the genetical control of the cell, will also be dealt with in more detail later.

Integration

Even from the incomplete account which has been given of the physico-chemical organisation of the cell it is clear that each particular region of the cell consists of a complex interlocking of very many simultaneously active physico-chemical systems. Each particular region of the cell has its properties defined by a vast group of variables, some of which are linked and some of which are independant. Our understanding of these is very far from complete. In some instances the necessary physics and chemistry is almost completely unknown. In very few instances are we able at present to deal quantitatively with these variables. When sufficient information is available to permit completely quantitative treatment, it is likely that the system will be so complex that it will be impossible to utilise this knowledge without the aid of electronic calculating machines.

REFERENCES

ADAM, N. K., 1941: *The Physics and Chemistry of Surfaces* (Oxford University Press, London).

BERNAL, J. D. and FANKUCHEN, A., 1937: *Nature* 139, 923.

BOURNE, G., 1950: *Cytology and Cell Physiology* (Clarendon Press, Oxford).

BRACHET, J., 1944: *Embryologie Chimique* (Masson, Paris).

CLARK, A. J., 1929: *The Mode of Action of Drugs on Cells.*

CLARK, A. J., 1937: *General Pharmacology.*

DANIELLI, J. F. and DAVIES, J. T., 1951: *Advances in Enzymology*, 11 (Academic Press, New York).

DARLINGTON, C. D., 1937: *Recent Advances in Cytology* (Churchill, London).

DAVSON, H. and DANIELLI, J. F., 1943: *Permeability of Natural Membranes* (Cambridge University Press, London).

GRAY, J., 1931: *Experimental Cytology* (Cambridge University Press, London).

MIRSKY, A. E. and RIS, H., 1948: *J. Gen. Physiol.*, 31,1.

STEDMAN, E. and STEDMAN, E., 1947: *Symposia Soc. for Exp. Biol. I.*

Symposia: *Cold Spring Harbour Symposia* (Darwin Press, New Bedford); *Society for Experimental Biology* (Cambridge University Press, London).
    1947: *Nucleic Acid.*
    1948: *Growth.*
    1949: *Selective Toxicity and Antibiotics.*

WILSON, E. B., 1928: *The Cell in Development and Heredity* (Mac Millan, New York).

Possible Actions of Drugs on Surfaces

Ionic Interactions

Ions may react with the charged groups of surfaces. The molecules composing a surface may have charges upon them which can be represented either as an electrostatic dipole, a fully ionised group, or as a combination of dipoles and fully ionised groups. When an interfacial layer of molecules is composed mainly of dipolar molecules, the electrostatic potential difference between the two phases is commonly of the order of 500 mV. Ions may affect the packing of these molecules if they come close to or are adsorbed upon the interface. The effect of the ions is, of course, in such a case mainly mediated by the interactions between the ionic charge and the dipoles of the molecules. Where fully ionised groups are present in the interface, such as carboxyl, phosphate and amino groups, the action of ions is mainly mediated by the interaction between the ion and the ionised group of the interface. The effect of ions upon an ionised surface is commonly greater than the effect of ions upon a surface composed of dipolar molecules. As a result of the interaction with ions the structure of a surface and its physical properties may be profoundly changed. For example, there may be large changes in viscosity. A monolayer of palmitic acid is a liquid, or even a gaseous film upon an alkaline solution of sodium chloride. But when the underlying cation is calcium, the monolayer is an extremely viscous liquid or a solid. The effects of ions upon monolayers can be readily studied by foll-

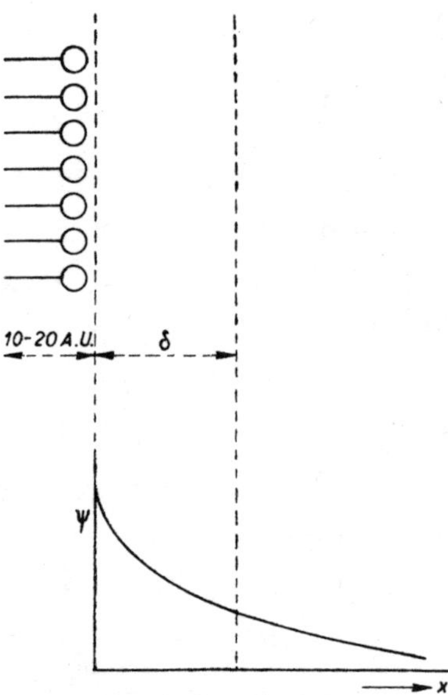

Fig. 4. Structure and thickness of an oil-water interface. The layer of oil molecules oriented at the interface is 10–20 A.U. thick, and (in solutions of uni-univalent electrolyte) the thickness of the electrical double layer δ is $3.1/\sqrt{c}$ A.U., where c is the electrolyte concentration. The lower part of the diagram indicates the variation in electrostatic potential ψ due to the charge on the surface, at various distances from the surface. At the point $x=\delta$, $\psi=\psi_0/e$ where e is the base of natural logarithms. The excess of ions at any point x due to the charge on the surface is proportional to $e\psi$: the greater part of the excess lies within the double layer of thickness δ.

owing the changes in viscosity of the surface film.

As a result of the forces operating between a charged surface and the ions present in an underlying medium, the ionic composition of the interfacial region may be very different from that of the surrounding bulk phase. To give an example: if a monolayer of palmitic acid is spread upon a solution containing 2,000 sodium ions to 1 calcium ion, in the interfacial region the ratio is of the order of 0.3 sodium ions to one calcium ion (DANIELLI and WEBB). This large difference in the ionic ratio is partly caused by the fact that the charge on the surface attracts multivalent ions much more strongly than it attracts univalent ions. But this is only part of the cause. In the case of the sodium palmitate monolayer just mentioned, if the electrostatic effect were entirely responsible for the difference in ratio between the surface and bulk phases, the ratio in the surface would be 100 sodium ions to 1 calcium ion: the difference between this figure and the actual figure of 0.3 sodium ions to 1 calcium ion is due to a second factor. The second factor is the formation of unionised complexes between certain ions and groups in the surface.

In the case of egg albumin molecules in solution it has been possible to carry this analysis to a quantitative conclusion (DANIELLI). The closed circles of Fig. 5, show experimental values obtained for sodium : calcium ratios in ovalbumin solutions. Ultrafiltrates were prepared from the solutions, and the difference between these ultrafiltrates and the composition of the initial solutions is due to the adsorption of ions at the surface of the ovalbumin molecules. The crosses of Fig. 5 show the results which would have been obtained if the excess of ions at the surfaces of protein molecules had been

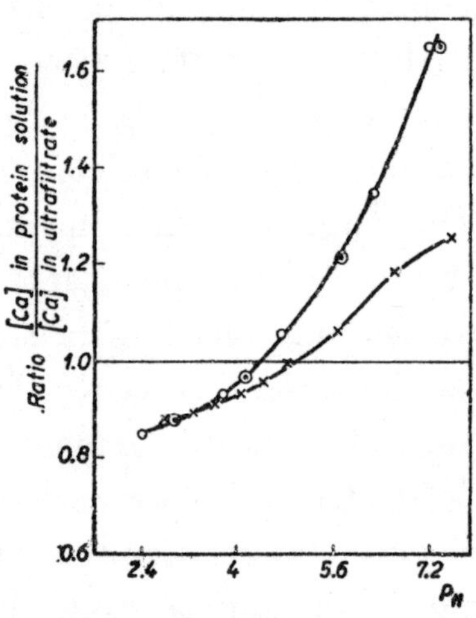

Fig. 5. The adsorption of calcium ions upon ovalbumin molecules. p_H is plotted horizontally, and the ratio: calcium in protein solution/ calcium in ultra filtrate, vertically. o experimental values; × calculated for electrostatic binding of calcium; ◎ calculated for formation of calcium-carboxyl complex with electrostatic binding

produced by electrostatic factors only: clearly only part of the effect can be attributed to electrostatic forces. The open circles of Fig. 5 are calculated on the assumption that the concentration of calcium at the surface of a protein molecule is raised by electrostatic interaction between the surface and the ions, and that the calcium ions enter into an equilibrium of the following type with the carboxyl groups present at the surface of the ovalbumin molecules.

$$Ca^{++} + R.CO_2^- \rightleftharpoons R.CO_2Ca^+ \tag{1}$$

It will be seen that the results obtained in this way are in excellent agreement with the experimental results.

I therefore suggest that in the mechanisms which we have just studied we have a rational approach to some aspects of the interaction of ions with surfaces. We shall now consider two examples of this.

The sodium/calcium ratio. It is very well known, indeed so well known that of recent years no explanation has been sought, that physiologically balanced salines, i.e. salines which will support the life of tissues in a relatively normal way, contain something of the order of 100 sodium ions to 1 calcium ion. It is difficult to conceive any physico-chemical mechanism which would

TABLE I

THE RATIO $[Na^+]:[Ca^{++}]$ IN PHYSIOLOGICAL FLUIDS, AND AT THE SURFACE OF CERTAIN CELLS

	Ratio in fluid medium	Ratio at surface
Red blood cell	50	2.6
Polymorphonuclear leucocyte	50	3.8
Arbacia egg	25	0.8
Astenas egg	25	1.2

permit 100 sodium ions to enter into a directly balanced relation with one calcium ion. But if the site of action of the ions is not in a bulk phase, but at surfaces such as the cell surface and the surface of protein molecules, then the position is radically changed. For example, if one calculates the ratio of sodium : calcium ions at the surfaces of cells in their normal environments, one finds that whereas the composition of the bulk phase surrounding the cells has a ratio of the order of 100 sodiums to 1 calcium, the ratio at the surface is of the order of 1 : 1 (see Table 1).

This then is a plausible theory for the explanation of the balanced action of sodium and calcium in physiological systems.

The Oligodynamic Effect of Heavy Metals. Let us suppose that the toxicity of a heavy metal is caused by the formation of unionised complexes between a surface such as that of a protein molecule and the metal ion. Then in the surface we have the following reaction:

$$Pr^- + M^+ \rightleftharpoons MPr \tag{2}$$

$$\text{i.e.} \quad \frac{[Pr^-][M^+]}{[MPr]} = K_M \tag{3}$$

From the conditions we have just stated it follows that equitoxic concentrations of different metals must give the same value for [MPr]. In equation (3) the terms

[Pr⁻] and [MPr] will be constant for equitoxic concentrations of the different metals. Consequently if K_M were known we could calculate the values of the equitoxic concentrations of the different metals, which are represented by [M⁺]. Now the equilibrium constant is

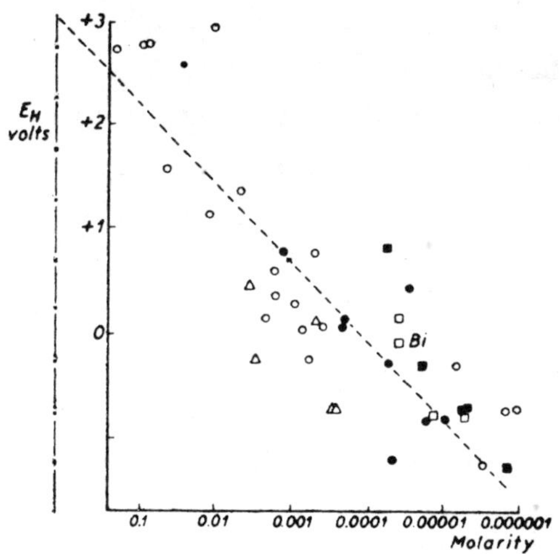

Fig. 6. The toxicity of heavy metals to various systems: The logarithms of the equitoxic concentrations of various metals are plotted against their standard electrode potentials. ■ 50 % inhibition of papain; □ 50 % inhibition of catalase; △ killing of *Paramecium;* ● killing of *Fundulus* egg; ○ killing of *Polycellis nigra* (planarian).

related to the standard electrode potential of the metal concerned (the reasons for this are given elsewhere, DANIELLI, 1946). If we substitute for K_M in equation (3) an appropriate term for the standard electrode potential E_M, and rearrange our equation we end up with

$$E_M = 0.058 \log [M] + \text{constant},\qquad(4)$$

i.e. if we plot log[M] against E_M we should obtain a

straight line. Fig. 6 shows the action of heavy metals on a number of biological systems. It is clear that the relationship we have just obtained is approximately true.

Thus we have at present a plausible theory of the oligodynamic action of heavy metals, and of the differences between their actions. It is an integral part of this theory that the action involves the formation of an unionised complex with ionogenic groups[1] at a surface. The types of ionogenic groups which we may particularly consider as likely to be involved are phosphate, carboxyl and SH. Much remains to be done before this theory can be regarded as established.

Dipole Interactions and Complex Formation

When molecules are present side by side in a monolayer the organisation of the monolayer is profoundly affected by the forces operating between adjacent molecules. If

---

[1] It is of interest to consider mechanisms whereby a charge upon a surface may arise. The mechanisms are four in number: 1. The partition of ions between the two phases; 2. The orientation of dipolar molecules at the interface; 3. The adsorption of ions; 4. The ionisation of ionogenic groups.

Mechanisms 2, 3, and 4 are obvious, but a word of explanation is required for mechanism 1. The different ions have different partition coefficients between say oil and water. But in order to preserve electrical neutrality the actual concentration of positive and negative charges in both bulk phases must be equal. This equality is brought about by the building up of an electrostatic potential at the interface which effectively modifies the partition coefficients of the different ions.

neighbouring molecules are ionised, the forces involved may be rather large, whereas the weakest forces encountered are Van der Waals' forces. Usually the forces which have to be considered can be ranked as follows: ion–ion > ion–dipole > dipole–dipole > Van der Waals' interactions. Since different molecular species have different distributions of attractive and repulsive forces in their molecules, at times the interaction between two adjacent molecules may be sufficiently selective to entitle us to regard the interaction as resulting in the formation of a two-dimensional molecular complex (SCHULMAN and RIDEAL). As a result of such complex formation, changes in at least five variables can be observed. These are: the packing of molecules, the surface pressure, the surface viscosity, the surface potential and the velocity of chemical reactions in the monolayer. The changes in these variables quite clearly are likely to be of importance as providing possible mechanisms of drug action.

Chemical reactions at interfaces have recently been examined, with the emergence of some interesting points. For example, the hydrolysis of trilaurin by hydroxyl ions has been shown to be sensitive to surface pressure changes. At a surface pressure of 5.4 dynes per cm the activation energy of the reaction is 10,000 calories, whereas at 16.2 dynes per cm the activation energy has risen to 15,000 calories. The actual orientation of a molecule at the interface may also exercise a profound effect upon the velocity of reaction. For example, with the simple esters, the rate of hydrolysis may vary, according to the configuration of the ester at the interface, so as to have a velocity as high as 0.18 min⁻¹, or as low as 0.005 min⁻¹ under a given set of conditions. Similar results have been attained for the rates of oxidation in interfaces of substances containing double bonds, by permanganate.

We see, from consideration of these results, that the interaction of an enzyme with its substrate may be affected by the orientation and other physical conditions of the substrate and that these in turn may be affected by drugs.

Specificity in Surface Reactions

Many drugs, such as adrenaline, veratrin, acetylcholine and cocaine act in such very low concentration that it is not at once apparent how a sufficient concentration of these molecules can arise at their site of action to produce a significant physico-chemical effect. However, if the action of these drugs is at surfaces it is possible to see both how an adequate concentration may arise, and why there is a high degree of specificity in their action. Each drug contains ionising groups, polar groups and practically non-polar hydrocarbon groups. Hence a drug has an intrinsic capability of being adsorbed at a surface, and when adsorbed, of altering the properties of the surface. E.g. let us consider the concentration of adrenaline, which may arise at a surface. This is given by Boltzmann's theorem,

$$\frac{C_s}{C_b} = e^{\frac{E}{RT}} \tag{5}$$

where C_s = concentration at the adsorbing surface; C_b = bulk concentration; E = energy of adsorption; R = = gas constant; T = absolute temperature.

The energy E may be regarded as made up of three components, one associated with the ionic groups, one with the polar groups and one with the non-polar groups. Minimal values for adrenaline are

Adrenaline	ionic	polar	non-polar	total E	C_s/C_b
	700	3000	3500	7200	2×10^5

Adrenaline:

$$\text{HO} \quad \text{OH}$$
$$\text{HO}\langle\text{ring}\rangle\text{—CH–CH}_2\text{–}\overset{+}{\text{N}}\text{H–CH}_3$$

The values taken for the energies are minimal. Yet when these energies are summed, the concentration of adrenaline which may arise at a surface is seen to be of the order of 10^5 times that which is found in the bulk phase. Under favourable conditions the energies may well be larger and could easily give rise to a relative concentration of the order of 10^{10}.

But before it is permissible to sum the energies in the way in which we have just done, it is necessary to provide an interface at which all three mechanisms for adsorption may become effective to an optimal degree. This means that the surface at which the drug is adsorbed must present an organisation of ionising groups, polar groups and non-polar groups as specific as that which is to be found in the drug itself. If this criterion is fulfilled the possible energy of adsorption is large. But a group in the wrong position, or having the wrong orientation, may readily prevent the dove-tailing of the drug and the surface and thus prevent many of the sites of potential adsorption becoming effective. Thus the presence of a methyl group in the wrong place may readily reduce the ease of adsorption of a drug by several thousand calories, and cause it to be relatively inert.

This theory may be regarded as a plausible one. It is very difficult to establish such theory.

Mechanism of Lysis

One of the aspects of drugs which has often attracted attention is the ability of certain substances to cause cytolysis. Cytolysis usually involves a direct action upon the plasma membrane of the cell. When lytic substances are considered from the point of view of surface chemistry it is seen that they are likely to produce their action partly by the formation of complexes with the molecules constituting the cell membrane, and partly by dissolving in the lipoid layer of the plasma membrane. As is indicated in Table II, some of these substances probably react primarily with proteins to form complexes, some react primarily with lipoids and some react with both.

There are a few substances, such as chloroform, which probably act mainly by dissolving in lipoids. There appear to be no instances of lysis which do not appear to be readily explicable in terms of the type of behaviour which is found at interfaces.

TABLE II

POSSIBLE MODES OF ACTION OF LYTIC SUBSTANCES

	React with proteins	React with lipoids	Dissolve in lipoids and denature proteins
Antibodies	+		
Trypsin	+		
Polyhydric phenols	+		
Heavy metals	+	+	
Soaps	+	+	
Digitonin		+	
Lysolecithin		+	
Lecithinase		+	
Bacterial toxins		+	
Chloroform			+
Phenol			+

The Action of Oestrogens on Monolayers

It is well known that many dihydroxy stilbenes have oestrogenic activity. SCHULMAN and RIDEAL have studied the action of these substances on protein monolayers, and have claimed that there is a maximum of oestrogenic activity coinciding with maximum activity on the protein monolayer (Fig. 7). This is an instance of a theory where there is an interesting correlation between a complex physiological effect and a simple physico-chemical phenomenon. As DALE has pointed out, this coincidence cannot be a sufficient explanation of the

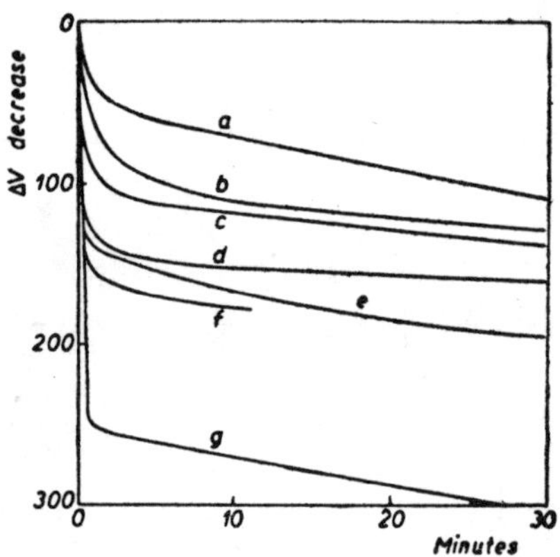

Fig. 7. The relationship between reduction of surface potential of protein monolayers and oestrogenic activity in the stilbene series. a. 4,4′ dihydroxy stilbene (1 : 150); b. 4,4′ dihydroxy dimethyl stilbene (1 : 40,000); c. 4,4′ dihydroxy ethyl stilbene (1 : 5,000); d. 4,4′ dihydroxy ethyl methyl stilbene (1 : 1,000,000); e. 4,4′ diethyl stilbene (1 : 3,000,000); f. 4,4′ dihydroxy hexadiene diphenyl (1 : 2,500,000); g. 4,4′ dihydroxy dipropyl stilbene (1 : 100,000). The reduction in surface potential increases steadily from a—f: substances e and f, which have the maximum oestrogenic activity, also combine most readily with the protein film. Substance g *replaces* the protein film.

physiological action of oestrogens in producing oestrus, for this change is restricted to certain types of cells of the body, whereas all cells are lavishly provided with protein monolayers. It is not, however, beyond the bounds of possibility that the action of oestrogens on monolayers is the basis, or part of the basis, of their general capacity to promote cell division. Very much more remains to be done, however, before these results of SCHULMAN and RIDEAL can be adequately claimed to be linked with physiology.

The Effect of Micelle Formation

In aqueous solution, when the properties of a solute are dominated by its polar groups, as is the case with methyl alcohol, the solute molecules are dispersed mainly as single molecules. But when there is a great excess of non-polar hydrocarbon groups in the molecule, practically speaking no single molecules exist in an aqueous solution of the substance concerned. Instead, the molecules are organised so that the non-polar parts are tucked away into micelles, which may be spherical or take the form of bimolecular sheets. The polar groups of the molecules are mainly to be found in the interface with the water, so that the non-polar part of the molecules does not in fact necessarily come into direct contact with water molecules. Substances forming solutions of this type are stearic acid, cholesterol and tripalmitin. Where there is a balance between polar and non-polar groups, as in $CH_3(CH_2)_9 \overset{+}{N} Me_3$, there is an equilibrium between single molecules and micelles, which is quite dynamic. If the surface tension of such a solution is studied as a function of concentration it is found that as the concentration is increased, at first the surface

tension falls very rapidly, until the concentration is reached at which micelle formation commences (Fig. 8). At this point, the curve rapidly flattens out, so that the surface tension becomes almost independent of concentration. Once the curve has become flat, the concentration of single molecules does not increase with increase

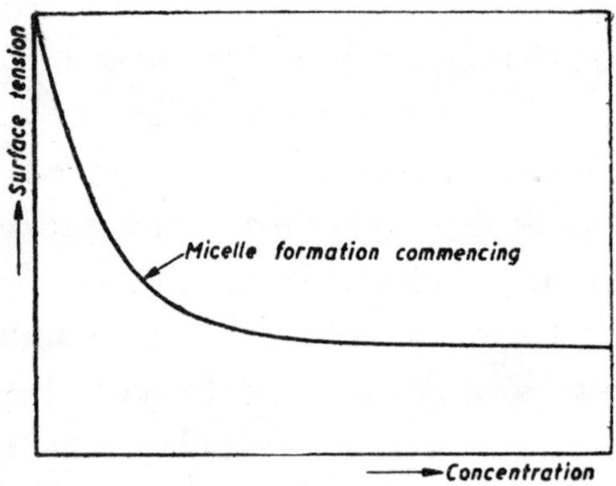

Fig. 8. The relationship between concentration, surface tension and micelle formation for a substance such as $CH_3(CH_2)_9\overset{+}{N}Me_3$

in concentration. What does increase is the concentration of micelles.

As a result of the possibility of micelle formation, the following consequences may occur.

1. If a substance can give rise to a micellar solution, and if the physiological action of the substance is a function of the concentration of single molecules in solution, then over the range of existence of micelles the action of a substance may be independent of its concentration. This is because over the range of existence of

micelles, the concentration of single molecules may be independent of the total concentration, increases in concentration resulting merely in formation of more micelles. The transition of micelles into single molecules of solute may have a very high temperature coefficient. Consequently the action of such a drug may also have a very high temperature coefficient.

2. If the activity of a homologous series of drugs is plotted against the number of carbons in the molecule, it is frequently found that somewhere in the region of nine carbon atoms a maximum of activity is reached. As the number of carbons in a series is increased, so also the ease of adsorption of the molecule increases, and consequently its activity per molecule rises. But as the number of carbons increases, so does the ease of formation of micelles, and it commonly happens with drugs containing an aliphatic carbon chain that the ease of formation of micelles increases more rapidly than the surface activity increases. Thus a point is reached at which micelle formation occurs before a concentration of the drug as single molecules is reached at which its physiological activity can become manifest. Hence the maximum in the curve of activity plotted against number of carbons (Fig. 9).

3. If a second substance is present, which can form micelles, a drug may be inactivated by combination with the micelles. This is illustrated by Fig. 10. In this diagram the rate of penetration of hexyl resorcinol into

Ascaris is plotted against the concentration of detergent present in the same medium. At first the penetration of the drug is increased by the presence of the detergent, but as the concentration of the detergent is increased a maximum is reached and the rate of penetration falls

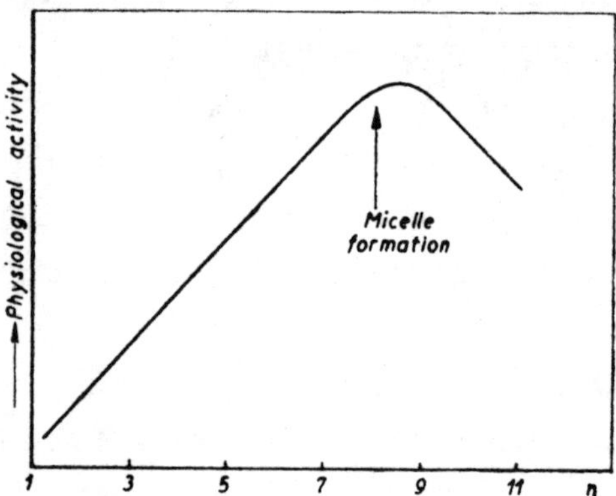

Fig. 9. The relationship between physiological activity, micelle formation and n, in a series $CH_3(CH_2)_nX$, where the physiological activity is a function of the concentration of single molecules

off practically to zero as the detergent concentration is still further increased (TRIM and ALEXANDER). In the same figure, the surface tension of the solutions is plotted also, and it will be seen that the point at which the rate of penetration begins to fall coincides roughly with the onset of micelle formation. It therefore seems very probable that the decline in rate of penetration of the drug is due to its forming a complex with the detergent when the latter is present as micelles, and that this complex is unable to penetrate the cuticle of *Ascaris*.

It is plain that the phenomenon of micelle formation enables us to furnish a plausible explanation of some of the peculiar properties of certain drugs. It is necessary

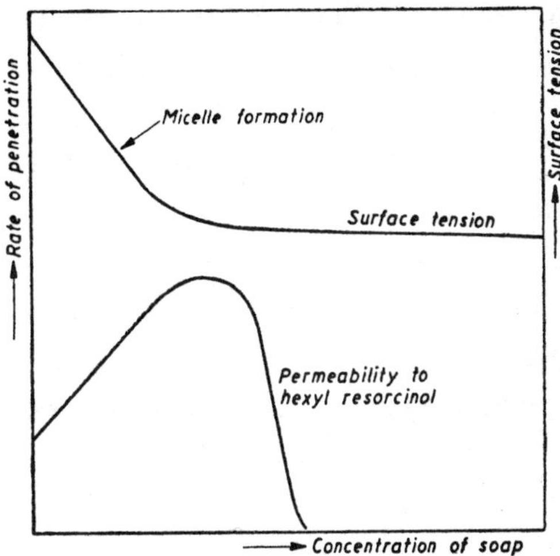

Fig. 10. The prevention of permeation of hexyl resorcinol into *Ascaris* by micelle formation in a soap solution (After TRIM and ALEXANDER)

to remark here that the complete establishment of these theories requires more than the demonstration of a coincidence between micelle formation and the onset of a change in the action of a drug.

Long-Range Forces

Recently the biologist has become interested in relatively long-range forces, i.e. forces operating over distances of the order of 25 A.U. up to several μ. BERNAL and FANKUCHEN have shown that in some types of colloidal solution molecules may be oriented parallel to one another by long-range forces. More recently ROTHEN has obtained results which can be interpreted to mean that relatively specific forces, such as those between antibodies and antigens, may extend over a distance of at least 100 A.U. The biologist is tempted by such phenomena as the adlineation of chromosomes in meiosis, and the reaction of cells to one another, to postulate similar forces extending up to several microns. If the field of action of these forces is as extended as some suppose, then they must be of fundamental importance in such phenomena as differentiation, chromosome mechanics, fibre adlineation and enzyme action. From the scanty information which is available it is already clear that the operation and the specificity of these forces is greatly affected by the net charge on the molecules concerned, and also by its detailed distribution. It is therefore plain that we have here a fertile field for the study of the action of drugs. But our knowledge is at present so restricted that it is not possible to do more than indicate the immense possibilities which exist here for future research.

REFERENCES

ABRAMSON, H. A., 1934: *Electrokinetic Phenomena* (Chemical Catalog Company, New York).

ADAM, N. K., 1941: *The Physics and Chemistry of Surfaces* (Oxford University Press, London).

BERNAL, J. D. and FANKUCHEN, A., 1937: *Nature*, **139**, 923.

DALE, H., 1943: *Trans. Farad. Soc.*, **39**, 320.

DANIELLI, J. F. and WEBB, D. A., 1940: *Nature*, **146**, 197.

DANIELLI, J. F., 1941: *Biochem. J.*, **35**, 470.

DANIELLI, J. F., 1944: *J. Exp. Biol.*, **20**, 167.

DANIELLI, J. F. and DAVIES, J. T., 1951: *Advances in Enzymology*, **11** (Academic Press, New York).

RIDEAL, E. K., 1945: *J. Chem. Soc.*, 423.

ROTHEN, A., 1947: *J. Biol. Chem.*, **168**, 89.

SCHULMAN, J. H. and RIDEAL, E. K., 1937: *Proc. Roy. Soc.*, B **122**, 29.

Symposia: 1949: *Surface Chemistry* (Butterworth, London).

1949: *Selective Toxicity and Antibodies* (Symposia Soc. for Exp. Biol., III).

TRIM, A. R. and ALEXANDER, A. E., 1946: *Proc. Roy. Soc.*, B **133**, 220.

Membrane Permeability and Drug Action

Introduction

There have been many academic studies of the permeability of natural membranes, but very few direct studies on permeability to drugs. Consequently, most of what can be said on this topic is based on permeability to molecules which are not usually regarded as drugs. In approaching this field we must distinguish between diffusion and secretion. Diffusion is the movement of molecules produced by thermal agitation; it now has a quantitative theory. Secretion is a process involving the expenditure of energy by a living organism to move molecules from one place to another: there are no quantitative theories of secretion.

The importance of permeability studies may be seen from consideration of the sites at which a drug may act. Even when some process such as absorption from the intestine is not involved, a drug always has to penetrate a cell membrane unless its action is on the external surface of a cell. In general, a drug which has penetrated a cell membrane may combine with a receptor group and also may be detoxicated more or less rapidly. For efficient drug action the rate at which a drug penetrates into a cell must be large compared with the rate of detoxication. Commonly drugs must also pass other membranes in addition to those of cells, such as the complex membranes composing the intestinal epithelium, the cuticle of a parasite etc. It is thus clear that there is much point

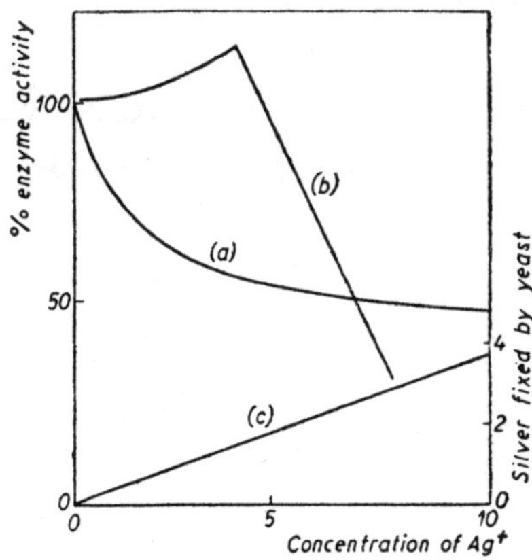

Fig. 11. The action of Ag on the inversion of sugar. (a) Inhibition of purified invertase; (b) inhibition of inversion by yeast cells; (c) amount of Ag taken up by yeast cells

in knowing how changes in the physical structure of a drug may modify its ability to permeate various types of membrane.

Even when we are studying the effect of drugs upon simple cell suspensions, such as those of bacteria and yeasts, the action may prove to be much more complicated than when we are dealing with those homogenates which are dear to the biochemists. Fig. 11 shows the action of silver on the inversion of sugar. Curve *a* shows that the action of silver upon free invertase is immediate, and always reduces the action of invertase. On the other hand, as is shown by curve *b*, when silver is added to a suspension of yeast cells, low concentrations of silver actually increase the rate of inversion of sugar, and it requires a relatively substantial concentration of sugar to destroy the enzyme action. It is possible that the first action of silver on the cells is to increase their permeability to sugar, and thus enable more substrate to obtain access to the enzyme than would otherwise be the case. When the concentration of silver is increased, the enzyme itself is affected. These results show that the permeability factor, even in a simple system, may be involved in a relatively complicated manner. In the following pages we shall to a large degree ignore complications and deal with permeability problems exclusively. But it is important to remember that in so doing we are indulging in an artificial abstraction.

Membrane Permeability and Drug Structure

When considering the permeation of drugs into cells, three major questions arise. These are:

1. Can the structure of the drug be modified without destroying its therapeutic activity?
2. Do different cells differ in permeability to the same substance?

3. Can the structure of a drug be modified to give predictable changes in permeability?

The answer to the first of these questions varies with the type of a drug which is under consideration. It is well known that the structure of the sulphonamides may

TABLE III

THE PERMEABILITY OF THE CELLS OF *Chara* TO DIFFERENT SUBSTANCES

	Living cell	Dead cell	Equal water cylinder
Methyl alcohol	1.3	0.8	0.27
Urea	320	0.9	0.34
Acetamide	24	1.2	0.38
Glycerol	1,700	1.9	0.49
Trimethyl citrate	5.5	2.2	0.67
Sucrose	50,000	4.1	0.92

The figures given are the times taken for the average concentration inside a cell to reach 50% of that outside. For comparison similar values are given for dead cells, and calculated values for a water cylinder of the same size as a *Chara* cell. (After COLLANDER)

be altered within very wide limits without destruction of activity. On the other hand among, say, the anti-malarial drugs, only comparatively small changes in structure are possible without loss of activity. Thus each group of substances must be considered as a separate case. So far as can be seen at the moment, unless the structure of a drug can be modified without change of activity, comparatively little can be done to make the best of the differences in permeability which may exist between the cells of the host and of the parasitic organism.

On the second question, as to whether cells differ in their permeabilities, there are two groups of data to be considered: the permeability of a given cell to different

TABLE IV

THE PERMEABILITY OF VARIOUS CELLS TO
DIFFERENT SUBSTANCES

	Ox red cell	Arbacia egg	Chara ceratophylla	Bact. paracoli	Gregarina sp.	Melosira sp.	Beggiatoa mirabilis
Trimethyl citrate	—	—	6.7	—	—	3.0	—
Glycol	0.2	0.7	1.2	—	0.7	0.4	1.4
Urea	8	—	0.1	0.08	0.3	0.04	1.6
Malonamide	—	—	0.004	0.03	—	0.02	—
Glycerol	0.002	0.005	0.02	0.06	0.02	0.03	1.06
Erythritol	—	—	0.001	0.005	—	0.01	0.8
Sucrose	—	—	0.0008	—	—	0.006	0.1

substances and the permeability of different cells to the same substance. Table III indicates very wide differences in permeability of the cells of *Chara* to different penetrating substances. It will be seen that the rate of permeation of the substances studied varies by a factor of 10^4. But the range of permeabilities to different substances is in fact even greater than this, for it is inconvenient to study the substances which penetrate extremely slowly. However, as in most instances substances which penetrate very slowly will not have a drug action, these more slowly penetrating substances are probably of little practical importance.

Table IV shows the permeability of a number of dif-

TABLE V

THE PERMEABILITY OF DIFFERENT CELLS TO UREA AND TO GLYCEROL

	Urea	Glycerol	Ratio urea/glycerol
Ox red cells	7.8	0.002	3,900
Chara ceratophylla	0.1	0.02	5
Plagiothecium denticulatum	0.004	0.003	1.33
Curcuma rebricaulis	0.002	0.002	1
Melosira sp.	0.04	0.03	1.33
Bacerium paracoli	0.08	0.06	1.33
Beggiatoa mirabilis	1.6	1.1	1.45
Gregarina sp.	0.7	0.02	35

ferent cells to various substances. In Table V some figures are given for the permeability of certain cells to urea and to glycerol. It will be seen from these results that the permeability of different cells to the same substance may vary by a factor of 100 or even 1000-fold. Moreover the relative order of permeation of substances into different cells is not always the same. Thus some cells are more permeable to urea than to glycerol, whereas others are more permeable to glycerol than to urea. From this it is apparent that considerable therapeutic advantages can in principle be derived if the drug is constructed to have permeability properties which resemble those of substances which readily penetrate into a parasitic cell, but which penetrate much less readily into the cells of the host.

Until recently, however, very few attempts have been made, and these unsuccessfully, deliberately to modify the structure of drugs so as to exploit the permeability characteristics of different types of cell. The main reason for this is that there was no general quantitative theory correlating permeability of cells with the structure of penetrating substances. Without such a quantitative theory it is really impossible to make much headway. But recently theoretical studies have been made which enable us to calculate the relative permeabilities of cells to different molecular structures, to a first approximation. It is convenient to divide the cases which are met in practice into five groups.

The first group is that where permeation occurs by bulk flow of fluid medium. As an instance of this we may take the formation of the glomerular filtrate in the kidney of mammals. Here all the crystalloical constituents of the blood are filtered off through a membrane, the pore size of which is very much larger than the diameter of a crystalloid molecule. In such cases all the molecules of a crystalloid character are carried along by the bulk flow of the fluid medium. Consequently the only force available to discriminate between different molecular types is the electrostatic force which may arise by the fact that the colloidal constituents of the blood are unable to pass the glomerular membrane, whereas the crystalloid ions which neutralise the charge on the colloids are able to pass the membrane. From this it follows that where a bulk flow occurs, as in an ultra-filtration, all molecules of the same charge type display the same permeability, provided they are small compared with the diameter of the pores through which the bulk flow occurs. All uncharged molecules have the same permeability. All univalent positive ions have the same permeability, but differ from the uncharged molecules. All univalent negative ions have the same permeability, but differ from the uncharged molecules and from the univalent positive ions, etc.

The second group is that in which permeation occurs by thermal diffusion through membranes the pores of which are large compared with the diameters of the diffusing molecules. In this case the molecules diffuse at different rates through the water filling the pores of the membrane, whereas in the previous case all the molecules were carried along by the bulk flow of the fluid in which they were dissolved. When diffusing through a membrane in this manner we find that the relationship

$$PM^{\frac{1}{2}} = \text{constant} \qquad (6)$$

is obeyed (P = permeability, M = molecular weight of diffusing molecule). Tables VI, VII and VIII show the permeability of collodion membranes, chitin and of the sulphur bacterium *Beggiatoa mirabilis*. Two instances of a collodion membrane are illustrated, one with a large pore size and the other with a small pore size. It will be seen that where pore size is large the permeability

TABLE VI

THE PERMEABILITY OF TWO COLLODION
MEMBRANES TO DIFFERENT MOLECULES

	Membrane (a)	Membrane (b)
Methyl alcohol	6.9	5.2
Ethyl alcohol	7.8	2.0
Propyl alcohol	7.7	0.8
Butyl alcohol	7.3	0.7
Ethylene glycol	6.3	0.2
Glycerol	7.8	0.2
Glucose	7.3	<0.05

The values are of $PM^{1/2}$. Membrane (a) is relatively permeable, and membrane (b) relatively impermeable

falls off inversely as the square root of the molecular weight, and equation (6) is obeyed to a first approximation. On the other hand, where the pore size is small, the permeability falls off much more rapidly than is indicated by equation (6). For membranes having a large pore size it is clear that to a first approximation the permeability may be calculated, provided that the permeability of the membrane to two or three other substances is already known and that the molecular weight of the drug is known. Table IX shows another interesting case, where equation (6) is obeyed to a first approximation: this is the diffusion of anions through the walls of the rumen of the sheep.

TABLE VII

THE PERMEABILITY OF CHITIN TO
DIFFERENT MOLECULES

	P	$PM^{1/2}$
Formic acid	19	127
Acetic acid	13.5	105
Propionic acid	14	122
Butyric acid	13	123

The two groups which we have just considered are those in which the molecule penetrating the membrane does not in fact leave the water in which it is dissolved. In the first case the permeating molecule was carried along by bulk flow of the water and in the second case thermal diffusion caused the permeating molecules to move through the water filling the pores in the membrane. The next two groups which we have to consider differ from the first two groups in that the permeating molecules actually pass from the aqueous phase in which they are dissolved into the non-aqueous phase constituting the membrane. In such cases the main resistance to permeation may lie either at the membrane-water interfaces or in the interior of the membrane. In the first case a diffusing molecule finds its passage through the membrane dominated by the difficulty of passing either from the water into the membrane, or from the membrane into

TABLE VIII

THE PERMEABILITY OF THE BACTERIUM
Beggiatoa mirabilis TO VARIOUS
MOLECULES

	P	$PM^{1/2}$
Glycol	1.4	11.0
Methylurea	1.2	10
Urea	1.6	12
Glycerol	1.1	10
Erythritol	0.84	9
Sucrose	0.14	2.5

TABLE IX

VALUES OF $PM^{1/2}$ FOR DIFFUSION OF ANIONS
THROUGH THE WALL OF THE RUMEN OF A
SHEEP

Acetate	5.8
Propionate	5.3
Butyrate	7.4

water. In the second case, passing the interfaces is not the limiting factor: the difficulty is in passing through the interior of the membrane. The group of cases in which passage through the cell membrane interface presents the main difficulty obeys the relationship

$$PM^{1/2}e^{\frac{2500x}{RT}}/B = \text{constant} \tag{7}$$

where B = oil : water partition coefficient, and x = number of unscreened CH_2 groups per molecule.

Table X shows the degree to which this relationship is

TABLE X

VALUES OF $PM^{1/2}e^{\frac{2500x}{RT}}$ FOR MOLECULES PENETRATING INTO THE CELLS OF *Chara ceratophylla*

Erythritol	8.5	Urotropin	5.3
Methylolurea	10.9	Methylurea	8.3
Urea	5.8	Dicyandiamide	9.4
Glycerol	6.5	Lactamide	6.4

obeyed by molecules penetrating the cells of *Chara cera- tophylla*. The adherence to the theoretical relationship is sufficiently good for one to be able to calculate the permeability to other molecules to a first approximation.

The fourth group of substances are those the main resistance to the penetration of which lies in the interior of the membrane. This group obeys the relationship

$$PM^{1/2}/B = \text{constant.} \tag{8}$$

Tables XI, XII and XIII show the degree to which this relationship is in fact obeyed for the cells of *Melosira, Arbacia* eggs and the rumen of the sheep.

Several comments must be made upon these results. First, the constants in the various equations just given are not the same for every cell type. A different constant is obtained for every type of cell, and the constant varies with species as well as with cell type. Thus for any particular cell, before the permeability to a drug can be calculated, it is necessary to know the permeability to

TABLE XI

VALUES OF $PM^{1/2}/B$ FOR VARIOUS SUBSTANCES PERMEATING INTO *Melosira* CELLS

Propionamide	7.2	Glycerol	4.5
Acetamide	2.0	Methylurea	2.3
Glycol	6.1	Urea	2.2

several other molecules. A second point which must be emphasised is that as drastic changes are made in the structure of the molecule, it is likely that one will pass from molecules falling into one of the groups given above to molecules falling into another of the groups. For example, group four consists mainly of molecules which permeate cells rather rapidly whereas group three consists of molecules which penetrate rather slowly.

TABLE XII

VALUES OF $PM^{1/2}/B$ FOR VARIOUS SUBSTANCES PERMEATING INTO *Arbacia* OVA

Butyramide	5.4	1:2 Dihydroxy-propane	1.8
Propionamide	5.9		
Acetamide	9.2	1:3 Dihydroxy-propane	2.8
Glycol	11.7		

With those cells which have a lipoid membrane determining their permeability, it commonly happens that as
the structure of the penetrating molecule is changed so
as to make it penetrate more slowly, a transference is
made from group four to group three. Another case

TABLE XIII

VALUES OF $PM^{1/2}/B$ FOR FATTY ACIDS PASSING THROUGH CELLS
OF THE SHEEP RUMEN

	P	$PM^{1/2}/B$
Acetic acid	3.5	2.7
Propionic acid	10.8	3.3
Butyric acid	27.6	3.4

which may be of practical importance is that of transference from group three to group two. This is illustrated by the results shown for the diffusion of substances through the sheep rumen. The sheep rumen
membrane consists of cells between which lies an inter-
cellular cement. Molecules such as the free fatty acids,
which can penetrate the cells rapidly, permeate the membrane mainly by passing through the cells, though, of
course, some free fatty acid also passes through the
pores. The fatty acids passing through the cells encounter
the main resistance to permeation in the interior of the
cell membranes they pass through, and thus fall into
group three. But the fatty acid anions permeate the cell
membranes much less readily. Consequently passage

through the pores is much more important in the case of the anions, and the anions to a first approximation fall into group two.

In the fifth group we collect together all those instances where substances, or groups of substances, do not conform to groups one to four or with the transition stages between the different groups. These are the instances in which secretory activity plays a part in determining the passage of the molecules across a membrane. In such cases, directly or indirectly, the cells concerned are expending energy to promote the passage of molecules across a membrane. Usually the permeability of cells in such cases is a non-linear function of the concentration of the molecule concerned. As the concentration of the molecules which are secreted increases in the cell environment, the secretory mechanism tends to become saturated, and the rate of secretion increases much more slowly than does the concentration in the environment. The existence of secretion is also often revealed by a degree of anoxia of the tissues concerned, or by the moderate use of enzyme poisons which interfere with the metabolic processes which provide the energy for secretion, or even in some cases possibly with the mechanism of secretion itself. At present our knowledge of secretion is much too slight for us to be able to predict with any confidence what the quantitative effect will be of a change in molecular structure upon the rate of secretion.

From the facts which have just been presented we may draw the general conclusion that, subject to certain reservations, the permeability of a cell to a drug can be calculated to a first approximation. E.g. when the structure of a drug is changed by adding or subtracting chemical groupings such as CH_2, $COOH$, or NH_2 etc., we can calculate the order of magnitude of the change in permeability which will ensue. In theory we can calculate the permeability of a membrane provided we know the details about its structure. Thus, for a porous membrane, we need to know the size of the pores, the thickness of the membrane and the numbers of the pores. With a lipoid membrane, we need to know the thickness of the membrane, the lipoid : water partition coefficients of the permeating molecule and the effective viscosity of the lipoid composing the membrane. Table xiv shows the permeability of several cells compared with the permeabilities calculated for a lipoid layer and for a water layer of the same thickness as that of the cell membrane. It will be seen that whereas the permeability of the lipoid membrane is of the same order of permeability as that of the cells, a water layer is about 10^9 times more permeable. Since there is no known method by which the viscosity of the interior of a cell membrane can be determined directly, in practice one has to determine it indirectly by determining the permeability to two or three substances. This, however, is in any case necessary before it is possible to decide to which group or groups the drugs in which we are interested belong.

The reservations mentioned above are of two kinds. If a substance happens to fall into one of the types which are secreted, we usually cannot calculate the effect of a change in structure. Then also there are small patches

TABLE XIV

THE PERMEABILITY OF VARIOUS CELL MEMBRANES TO CERTAIN SUBSTANCES, COMPARED WITH THE PERMEABILITY CALCULATED FOR AN EQUIVALENT THICKNESS OF WATER, AND OF OIL HAVING A VISCOSITY 100 TIMES THAT OF WATER

	Arbacia egg	*Chara cerato-phylla*	*Plagio-thecium denticu-latum*	5 mμ oil	5 mμ water
Propionamide	2.3	3.6	0.2	3.0	1.4 $\times$ 10^9
Acetamide	1.0	1.5	0.7	0.8	1.8 $\times$ 10^9
Glycol	0.73	1.2	0.3	0.7	1.7 $\times$ 10^9
Urea	—	0.11	0.004	0.02	1.8 $\times$ 10^9
Glycerol	0.005	0.02	0.0003	0.005	1.4 $\times$ 10^9
Malonamide	—	0.004	0.0008	0.002	1.4 $\times$ 10^9
Erythritol	—	0.001	0.00007	0.00007	1.2 $\times$ 10^9

on the surfaces of at least some cells which appear to be specially adapted to permit rapid permeation by a given molecular species without permitting a similar advantage to other molecular species. These patches may, of course, be part of the cellular secretory mechanisms, but until we know more about them it is wiser to assume that they may be distinct from active secretory processes. The fact that these reservations must be made should not be taken as indicating that there need be any hesitation in applying the conclusions derived from studies of cell permeability to practical problems. What is necessary in such circumstances is to keep an alert eye for the complications which may arise from failure to obey the laws of thermal diffusion across membranes.

Problems of the Access of Drugs to Organs

The permeability problems involved in the study of drug action are far from limited to those encountered in the study of the permeability of cell membranes. Frequently, the problem of access to an organ, or of absorption from the digestive tract, constitutes a more important difficulty than permeation into cells of either the host or the parasite. The main practical problems which tend to arise are: 1. inadequate adsorption from the digestive tract, 2. peculiar permeability properties which prevent a drug reaching a particular organ, e.g. the brain, 3. securing an effective concentration in one organ may involve a toxic concentration elsewhere.

From these difficulties there are at present three general procedures to which resort may be made. The first of these is to modify the rate at which the drug may permeate an organ by simple diffusion. Thus if a drug is needed to penetrate the central nervous system, an increase in its lipoid solubility should be sought, whereas if permeation of the central nervous system is an undesirable feature, increasing the polar character of the drug, for example by its administration as a glucoside, is suggested. The second possibility is to modify the structure of the drug so as to increase the probability that it will, or will not, fit into the secretory pattern of certain organs. The combination of a drug with cholic acid may secure a heavy secretion by the hepatic cells. Increasing the polar-nonpolar asymmetry of the structure of a drug is very likely to increase its secretion by the kidney. It may also secure its penetration into the central nervous system, and there are indications that permeation into the mammary gland may be favoured in this way. A thorough study of the so-called blood-brain barrier from this point of view is likely to lead to valuable results, and the economic problem of mastitis in cattle might well yield to a similar study of the secretory activity of the mammary gland. A third possible mechanism is to administer a drug so that it shall be inactive except at selected sites of action. This mechanism has been used, for example, for drugs involving the grouping ⟨ ⟩As=As⟨ ⟩. Compounds of this type are able to penetrate into the central nervous system, whereas the simple arsenoxides R⟨ ⟩AsO cannot do so readily. To get a therapeutic concentration of arsenoxide in the central nervous system is likely to involve a toxic concentration elsewhere. But if the drug is given as R_1⟨ ⟩As=As⟨ ⟩R_2 it penetrates into the central nervous system relatively readily, and is there transformed into the therapeutically effective arsenoxide. There are probably many instances where such procedures could be adopted intentionally, instead of by chance, as was the case with the arsenic compounds. Thus a drug administered as a phosphate ester would be likely to be inactive until the phosphate had been hydrolysed away from the rest of the drug. Thus such a compound might well display a high activity only at sites having a relatively high concentration of phosphatase, such as the kidney and bones. If it is true, as has been suggested recently, that tumours have a higher concentration of glucuronidase than normal tissues, it is possible that a drug relatively selective in its action could be obtained by administration of a toxic substance as a glucuronide. In the same way, since the cells of tumours of the prostate are commonly rich in acid phosphatase, a phosphate ester might increase the specificity with which a drug can act upon this tumour. Instances of this sort could be multiplied indefinitely.

Optimal results are likely to be obtained in any individual instance by combining several of the devices mentioned above. For example, if one wishes to obtain a drug which will penetrate relatively well into the central nervous system, one would tend to study the effect of increasing its lipoid solubility, increasing its basic character and increasing its polar-nonpolar asymmetry.

Examples of the Permeability Factor in Drug Action

To conclude this chapter I shall give three examples in which the importance of the permeability factor in the study of drug action is readily made apparent from the practical point of view.

Drugs acting on Ascaris. A number of substances are known to be of practical value as anthelminthics. These include thymol and hexyl resorcinol. But resorcinol

TABLE XV

EXPERIMENTAL AND CALCULATED VALUES OF
THE PERMEABILITY OF *Ascaris* CUTICLE

	P observed	P calculated
Resorcinol	0.7	0.05
Butyl resorcinol	3.0	(3.0)
Hexyl resorcinol	15	26
Heptyl resorcinol	37	80
Thymol	10	11
Chloroform	20	7
Nicotine	0.1	0.1

itself is of no practical value. TRIM has recently studied the permeability of *Ascaris* to various drugs. The results are shown in Table xv. The rate of permeation into *Ascaris* obeys the relationship

$$PM^{1/2}/B = 0.2 \text{ approximately,}$$

i.e. the membrane controlling the permeation of substances into *Ascaris* belongs to group three of the types of permeation given on p. 52. The permeability of *Ascaris* cuticle to a drug can be readily calculated from the equation $P = 0.2B/M.^{1/2}$ The figures obtained for the permeability are sufficiently close to the experimental ones to indicate the relative likelihood of various substances permeating in concentrations sufficient to have a toxic action. For example, the results show that the active substance hexyl resorcinol penetrates into *Ascaris* much more rapidly than does the inactive substance resorcinol.

The Toxicity of Arsenoxides to Trypanosomes. The toxicity of arsenoxides to trypanosomes has been the subject of an intensive study by a number of investigators, particularly KING and HAWKING. A great number of arsenoxides have been synthesised and their toxicities determined by finding the lethal dilution (L.D.), i.e. the number of litres in which 1 gram mol. of each substance must be dissolved to obtain a given degree of killing of trypanosomes in a given time. The substances concerned are of a type which we should expect to permeate trypanosomes readily, and to conform to group four of the permeation groups given on p. 52. The dilutions in which these drugs are effective are very high, and it seems likely from the studies of HAWKING in particular that practically every molecule of arsenoxide which permeates into a trypanosome becomes fixed by a receptor grouping such as SH. Thus a trypanosome is killed when a given proportion of its receptors is saturated, i.e. when a given amount of arsenoxide,

TABLE XVI

VALUES OF THE LETHAL DILUTION (L.D.) FOR THE KILLING OF TRYPANOSOMES BY CERTAIN ARSENOXIDES, AND CALCULATED VALUES OF $(\text{L.D.})M^{1/2}/B$

	L.D.	$(\text{L.D.})M^{1/2}/B$
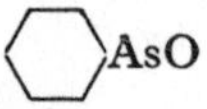 ⬡AsO	1900	2500
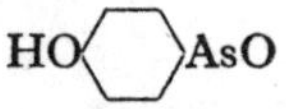 HO⬡AsO	530	6900
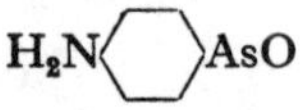 H_2N⬡AsO	33	580
 $CO_2H.CH_2.NH$⬡AsO	3.8	5400
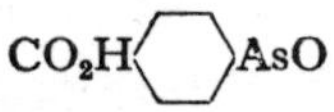 CO_2H⬡AsO	0.39	2000
Variation	5000 fold	12 fold

The relatively small variability of $(\text{L.D.})M^{1/2}/B$ indicates that most of the variation in L.D. is caused by differences in the rates of permeation of the arsenoxides into the trypanosome.

irrespective of its structure, has penetrated into the trypanosome. If this is true, the permeability of the trypanosome must be the main factor determining the value of the lethal dilution. Assuming that this is so, and that the permeation group is group four, it is easy to show that the following relationship should hold:

$$\frac{(L.D.)M^{\frac{1}{2}}}{B} = \text{constant approximately.}$$

Table XVI shows a small selection of the experimental data, including the most toxic, the least toxic and several intermediate compounds of the arsenoxide type. Whereas the value of the lethal dilution varies by a factor of 5,000 fold, the value of $(L.D)M^{\frac{1}{2}}/B$ varies by only twelve fold, i.e. of the variation in the L.D., practically the whole is accounted for by the variation in permeability. Thus the organic chemists, in synthesising a wide range of arsenoxides, were unwittingly studying the permeability of trypanosomes and little else.

The Chemotherapy of Lewisite Poisoning. In a series of biochemical studies PETERS, STOCKEN and THOMPSON found evidence that lewisite exercises its toxicity by combining with the SH groups of enzymes. By studying the ease with which lewisite may be detached from the compound it forms with kerateine, they concluded that lewisite must often combine not merely with one, but with two thiol groups of the protein molecule. Consequently they argued that, to obtain a substance which would compete efficiently with tissue enzymes for lewisite, and thus constitute an efficient therapeutic reagent, it would be necessary to have a dithiol grouping. VOEGTLIN and QUASTEL had both shown earlier that monothiols can prevent or reverse some of the toxic action of arsenoxides.

The danger from lewisite usually arises from skin contamination. Consequently to prevent vesication by lewisite, it is necessary to have a dithiol which will readily penetrate the skin. To attain reasonably fast permeation of skin, it is necessary to have a small molecule, having a moderate oil-water partition coefficient. Glycerol α, β-dithiol, later known as B.A.L.[1] conforms to this specification and was found to be very effective as an antidote to the vesicant action of lewisite.

B.A.L. also has value as an agent for systemic arsenical poisoning. But this is limited by the toxicity of B.A.L. itself. The maximum dose of B.A.L. which may be administered to man is 4 mg/kg/4 hrs. This is considerably below the amount of dithiol which would be needed to secure efficient therapeutic action in a serious case of systemic lewisite poisoning. Consequently an attempt was made to obtain a substance which would be an efficient antidote for systemic poisoning. The points in the molecular specification for such a substance were:

1. It should inactivate arsenic.
2. It should have a low toxicity.
3. It should remove arsenic from cells into which arsenic has penetrated.
4. It should penetrate the whole of the lymph and vascular spaces.
5. The complex formed with arsenic should be readily excreted.

[1] B.A.L. for British Anti-Lewisite.

In chemical terms these specifications can be met in the following ways. A dithiol grouping in the molecule will provide for point 1. Increasing the polarity of the compound will comply with point 2 by reducing the rate of penetration into cells: to increase the polarity it is necessary to increase the number of OH, COOH or SO_3H groups in the molecule. Point 3 can be met by having a relatively high concentration of the dithiol in the blood stream: intracellular arsenic is not completely incapable of diffusing out of cells, and if there is a high concentration of external dithiol, every arsenic molecule which diffuses out of a cell will be trapped by the dithiol. To meet requirement 4 it is necessary that the molecule should not be very large: for example, thiostarch or thioglycogen are excluded by this condition. The last requirement, ready excretion, is achieved by the same means as requirement 2.

Thus the specification of the molecule appeared as follows:

$$
\begin{matrix}
\text{C.SH} \\
| \\
\text{C.SH}
\end{matrix}
\quad + \quad
\left.
\begin{matrix}
2\ \text{COOH} \\
3\ \text{OH} \\
\text{or}\ 1\ SO_3H
\end{matrix}
\right\}
\quad \text{as alternatives}
$$

With a specification as precise as this it was not necessary to investigate many compounds. In fact, only two dithiols were studied, dithioadipic acid and the glucoside of B.A.L. The glucoside of B.A.L. was the better, having an $L.D._{50}$ of the order of 7.5 g/kg. When administered to rabbits which had received an $L.D._{95}$ of lewisite, the glucoside would ensure 100% survival with treatment begun not less than 4 hours after contamination and 50% survival with treatment begun not less than 6½ hours after contamination. Untreated animals die in about 12 hours.

In the first experiments with this compound, point 3 of the specification was assumed to be sufficiently met by the diffusibility of the arsenic itself, which would cause the arsenic to be trapped by the dithiol in the blood stream. Later an attempt was made to increase the diffusibility of the arsenic by administering small amounts of B.A.L. This can penetrate readily into cells and there combines with the arsenic which was hitherto combined with intracellular proteins. It was thought that as a result of this process arsenic would diffuse much more readily into the blood stream and would there be trapped by the large concentration of B.A.L. glucoside. It was in fact found that amounts of B.A.L. which were too small to affect the percentage of survival when given alone, and which were too small to exercise a toxic effect, would markedly increase the efficiency of the glucoside.

It will be apparent from what has been said in this chapter that, even with our present limited knowledge of the permeability of mammalian tissues, it should often be possible to reach valuable conclusions about the action of a drug and the design of new drugs. In particular it is often possible to decide whether the permeability factor is limiting the effectiveness of a drug, and to decide what changes in permeability would be profitable. The systematic employment of this information should prevent much waste of time in the synthesis and testing of drugs. The great weakness in our present understanding in this field is our limited knowledge of secretory processes.

REFERENCES

CLARK, A. J., 1937: *General Pharmacology.*

COLLANDER, R., 1937: *Trans. Faraday Soc.*, 33, 985.

COLLANDER, R. and BARLUND, A., 1933: *Actu. Bot. Fenn.*, 11, 1.

DANIELLI, J. F., McANALLY, M. and PHILLIPSON, J., 1946: *J. Exp. Biol.*, 20, 417.

DANIELLI, J. F. and others, 1947: *Biochem. J.*, 41, 325.

DAVSON, H., 1940: *J. Cell. Comp. Physiol.*, 15, 317.

DAVSON, H. and DANIELLI, J. F., 1943: *The Permeability of Natural Membranes* (Cambridge Press, London).

LILLIE, R. S., 1923: *Protoplasmic Action and Nervous Action* (Chicago).

PETERS, R. A., 1947: *Nature*, 159, 149.

PETERS, R. A., STOCKEN, J. R. and THOMPSON, R. H. S., 1945: *Nature*, 156, 616.

QUASTEL, J. H., 1947: *Nature*, 159, 824.

Symposia: Faraday Society, 1937: *Permeability*, Trans. Faraday Soc;
1940: *Mode of Action of Drugs*, Trans. Faraday Soc.
Society for Experimental Biology, 1949: *Selective Toxicity and Antibiotics.*

VOEGTLIN, F. R., 1925: *Physiological Reviews*, 5, 63.

WINTERSTEIN, H., 1926: *Die Narkose* (Berlin).

CHAPTER IV

Enzymes and Drug Action

Functions of Enzymes

In recent years many biologists have emphasised the likelihood that a large part of the action of drugs upon cells is to be explained mainly by the action of the drugs on cellular enzymes. In England, for example, this was particularly emphasised by A.J.CLARK, R.A.PETERS, and D.KEILIN. It may not be immediately apparent why this should be so, but reasons become clear enough if we consider the functions of enzyme systems in cells. These functions include 1. the synthesis of substances which act as an immediate source of potential energy for the physiological activity of the cell, e.g. the synthesis of adenosine triphosphate; 2. the conversion of potential energy to mechanical work, as is seen in muscular contraction; 3. the protection of the cell against invasion by foreign bodies—thus foreign proteins are destroyed by proteases, *d*-amino acids by *d*-amino oxidase, hydrogen peroxide by catalase; 4. secretory activity is dependent, directly or indirectly, upon enzyme activity; 5. evidence has arisen, both from cytochemical studies and from the study of mutations of yeasts, moulds and bacteria, that enzymes may be concerned in the mediation of genetic effects. In fact, there are singularly few activities of living cells in which one of the key positions is not occupied by one or more enzyme systems.

Possible Functions of Drugs in Relation to Enzymes

Accepting the fact that enzymes are of vital importance in the activity of cells, we must now consider the various ways in which the activity of enzymes may be modified by a drug. The possible modes of activity are quite numerous. They include the following.

1. *Action as carriers.* Substances such as methylene blue and pyocyanin may act as carriers between atmospheric oxygen and dehydrogenases: in so doing the normal carrier systems, such as cytochrome, are short-circuited. Such an action may not at first sight appear to have serious consequences. But in practice the consequences may be quite dramatic: for example, sea urchin eggs have their respiration raised by about 200% by addition of pyocyanin, and this rise is accompanied by an almost complete cessation of the processes of cell division.

2. *Action as activators.* Some substances are able to modify the structure of an enzyme and thus modify its activity. For example, reducing agents, such as B.A.L. and HCN, are able to activate SH enzymes.

3. *Action as chemical inhibitors.* A number of sub-

stances such as the arsenoxides, the nitrogen mustards, iodoacetate, fluoride and iodine, are able to inhibit the action of enzymes by forming a chemical compound with chemical groups which are essential for the maintenance of enzyme activity.

4. *Action as physical (competitive) inhibitors.* In this category we may mention malonate, which acts as an inhibitor for succinic dehydrogenase; glyceraldehyde, which acts as an inhibitor for triose phosphate dehydrogenase, and the sulphonamides, which are believed to compete with p-aminobenzoic acid for enzyme systems concerned in the metabolism of the latter.

5. *Action as prosthetic groups.* Certain substances are able to act as the prosthetic groups of enzymes, thus activating previously existing apoenzyme molecules. Examples are vitamin B_1, which is concerned as a prosthetic group in the pyruvic oxidase system, pyridoxal, which is a prosthetic group for some decarboxylases, and vitamin B_2, which is a constituent of flavoprotein systems.

6. *Action as coenzymes.* As an example of this may be mentioned nicotinic acid, which is incorporated into the molecule of coenzyme II.

7. *Action as cosubstrates.* No examples of this are known, but abnormal cosubstrate activity could clearly be a serious source of trouble to a cell. It is but comparatively recently that BERGMAN introduced the conception of cosubstrates, so that it is not surprising that

no examples of such activity have yet come to light[1].

8. *Action as substrate removers.* In the living cell the course of metabolism of a particular substance is in part determined by the presence of a suitable chain of enzyme systems. The functioning of such a chain of enzyme systems is dependent upon the product of the action of one enzyme passing on to another enzyme for which it is a specific substrate. Serious interference may occur by modification of the structure of an intermediate, in the course of its passage by diffusion from one enzyme to another. For example, the chain of enzymes concerned in the anaerobic metabolism of glucose may have their action disrupted in this way by HCN, or by H_3AsO_4. The action of HCN is to form a cyanhydrin with phosphoglyceraldehyde, thus removing the substrate for phosphoglyceraldehyde dehydrogenase. H_3AsO_4 excercises its toxicity in part by combining with phosphoglyceraldehyde under the action of triose phosphate dehydrogenase, so that as a result of the activity of this enzyme a phosphoarsenoglyceric acid results instead of diphosphoglyceric acid. The phospho-

[1] Cosubstrate activity is best understood by considering an example. When trypsin is added to a solution of glycyl-leucine, the dipeptide is not split by the enzyme. But if, now, to the solution is added a little acetyl-phenyl-alanyl-glycine, synthesis occurs of a little of the substance acetyl-phenyl-alanyl-glycyl-glycyl-leucine. Following this synthesis, trypsin splits off first leucine and then glycine, leaving acetyl-phenyl-alanyl-glycine as a residue which is not attacked. The compound acetyl-phenyl-alanyl-glycine is said to have cosubstrate activity in the splitting of glycyl-leucine by trypsin.

arsenoglyceric acid decomposes spontaneously, so that no diphosphoglyceric acid is available as substrate for the next enzyme in the series.

Problems in the Analysis of the Action of Drugs on Enzymes

When a drug is acting upon a simple solution of an enzyme, the analysis of the effect of the drug may be relatively simple. For example, when the percentage inhibition of an enzyme is plotted against the logarithm of the concentration of the inhibiting drug, a linear curve is quite commonly obtained. When drugs are acting

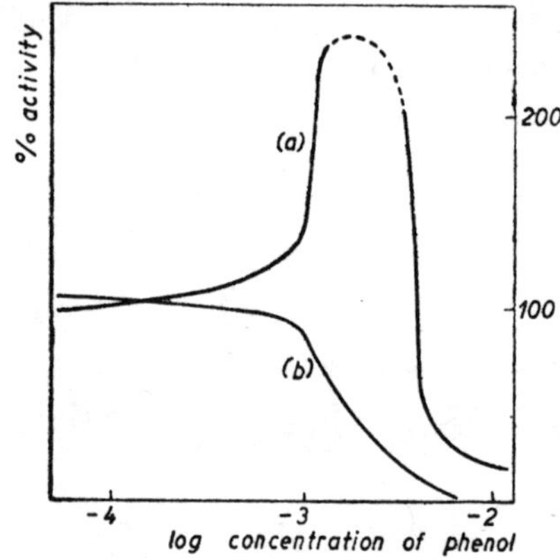

Fig. 12. The respiration of yeast (a) and the fermentation of sugar by yeast (b) as affected by phenol

upon cells, the action-concentrations curves are sometimes linear, sometimes non-linear. And even when they are linear it does not follow necessarily that the action of the drug on a given enzyme system is the same as it would be on the same enzyme in aqueous solution. The analysis is further complicated by the fact that the

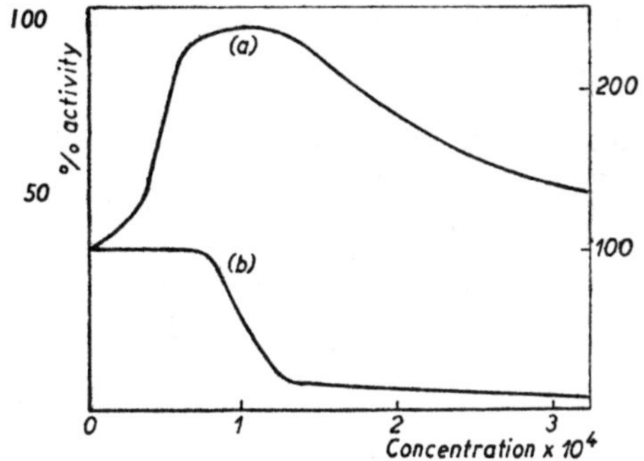

Fig. 13. The effect of dichlorophenol upon the respiration (a) and cleavage (b) of *Arbacia* ova. Note that there is little effect on cleavage until the increase in respiration is almost complete.

effect of an enzyme poison may affect different cellular processes in different ways. As is shown by Fig. 12 phenol increases the rate of fermentation of sugar by yeasts, but decreases the rate of respiration. Fig 13 shows

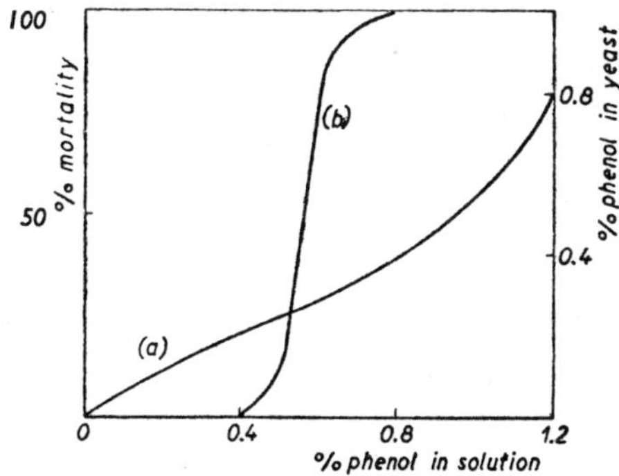

Fig. 14. The relationship between the uptake of phenol by yeast (a), and the lethal action of phenol on yeast (b)

that whilst dichlorophenol increases the rate of respiration of *Arbacia* eggs, it decreases their rate of cleavage. In such instances, when there are two or more cellular processes which are interfered with in different ways by the same drug, which is to be taken as the index of activity on cellular enzymes? There is no simple answer to this question. Furthermore, if we endeavour to correlate the

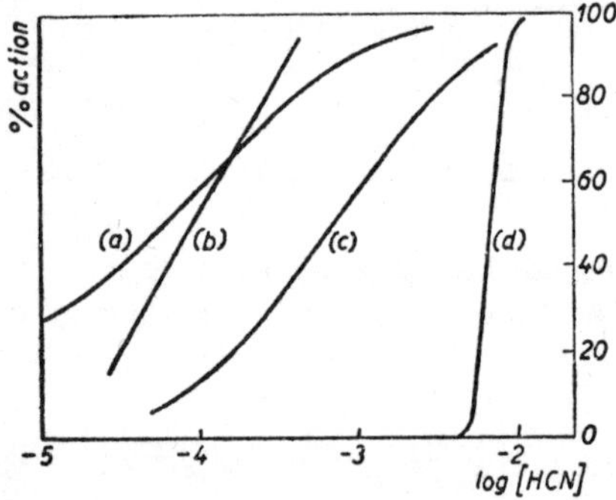

Fig. 15. The relationship between concentration of HCN and inhibition of various physiological processes. (a) Assimilation of CO_2 by *Chlorella*; (b) oxygen comsumption of frog's ventricle; (c) mechanical response of frog's ventricle; (d) lethal action on *Tribolium confusum*

amount of drug taken up with the change in a particular activity, it is commonly found that there is no close correlation. Fig. 14 shows, for example, that when phenol is acting on yeast there may be a considerable uptake of phenol before a significant change in physiological activity is observed. The first moiety of the phenol taken up appears to be inactive.

It is thus clear that the action of drugs on intracellular enzymes must be difficult to analyse.

The analysis is further complicated by the occurrence of very marked species differences. For example, when we take the inhibitory action of HCN on cellular processes or say the lethal action of H_2S, it is tempting to work on the hypothesis that in all cases the drug is acting on the same enzyme system, but Figs. 15 and 16 show that the concentrations required to produce a given degree of

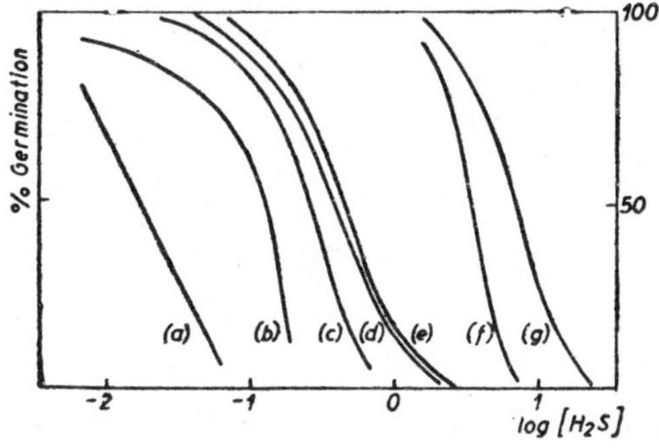

Fig. 16. The relationship between concentration of H_2S and its lethal action on the spores of eight different species of fungi. (a) *Venturia inequalis* and *Uromyces caryophyllinus;* (b) *Puccinia antirhini;* (c) *Sclerotina americana;* (d) *Macrosporidium sarcinaeforme;* (e) *Pestolatia stellata;* (f) *Glomerella cingulata*; (g) *Botrytis*

action vary widely from species to species. Various explanations of this are possible. It may be that the drug does not produce its effect by acting on the same enzyme in all species. Or it may be that a given enzyme in different species varies in its susceptibility to a given drug. A third possibility is that the effective concentration of the drug which arises in the vicinity of the enzyme is different in different cells, although the external concentration of drug is the same.

The Action of Drugs on Respiration and Glycolysis in Muscle

From the evidence which can be derived from studies on living cells it is clear, as we have just seen, that there are great difficulties in analysing the action of a drug into terms of activities on specific enzyme systems. An alternative approach to the problem can be made by studying the enzyme systems involved in a particular physiological process on the test-tube scale, i.e. using tissue extracts. This procedure has been carried out in great detail in the case of the respiration and glycolysis of muscle. The results so obtained seem likely to be representative of the type of conclusion which will be reached when a completely satisfactory analysis of drug actions on enzymes is available. At present we must have some reservations about the theories put forward by the biochemists, because the enzyme systems have in most cases been shown to perform the functions ascribed to them *in vitro* only. In the case of the cytochrome system, KEILIN and others have provided direct evidence that the intracellular systems are behaving in the way which is postulated from test tube experiments. But so far as glycolysis is concerned, whilst the picture built up by biochemical studies is very plausible, we still await conclusive evidence from studies on living tissues that the chain of events is identical with that postulated.

With these, and certain other reservations which will

be mentioned later, we may proceed to examine the action of drugs on the respiration and glycolysis of muscle.

Fig. 17 gives a rough outline of the main steps involving the use of oxygen and glucose by muscle. Also included are the probable points of action of carbon mon-

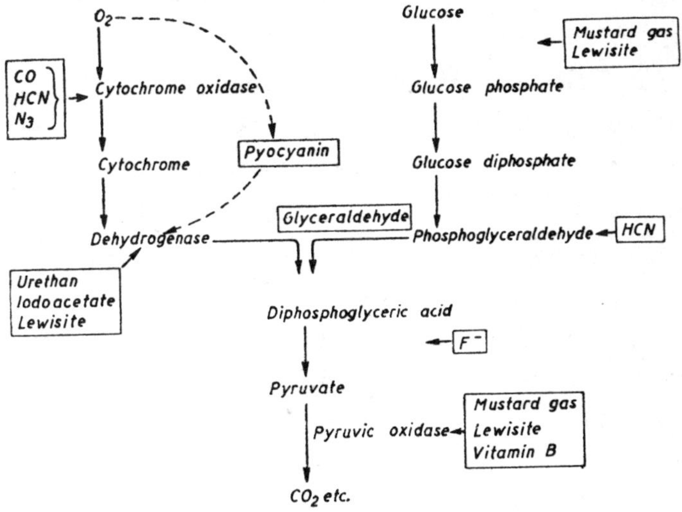

Fig. 17. A condensed scheme of the enzyme-catalysed steps in the respiration and glycolysis of muscle, with the points of action of some drugs

oxide, hydrocyanic acid, azide, urethan, iodoacetate, lewisite, mustard gas, glyceraldehyde, arsenic acid, fluoride and vitamin B_1. As is indicated in the figure, we have strong reasons for believing that the enzyme systems concerned in respiration and glycolysis are organised to give rise to a chain of reactions in which each reaction produces a reaction product which is the substrate for the subsequent reaction. As is indicated

by the diagram, there is a strong tendency for each drug
to act rather selectively on a particular enzyme system.
But this is no more than a tendency, and with the enzyme
systems with which we are concerned here some of the
drugs act at several points. For example, HCN can act
both on the cytochrome system and also by combining
with phosphoglyceraldehyde: lewisite can act on hexokinase, on triose phosphate dehydrogenase and on the
pyruvic oxidase system: mustard gas can also act at more
than one point. From these observations it is clear
that a drug may have the potentiality of acting at several
stages, even in one chain of biochemical events. Consequently the analysis of the action of a drug on the intracellular enzymes is extremely complicated. In the system which we have been considering there are about
twenty enzyme systems only. In the living cell there
must be thousands of enzyme systems. To decide which
of these enzyme systems is of importance in the mediation of the effect of any particular drug is obviously a
task of the first magnitude. In a few cases, where the
action of the drug is very rapid, as is the case with HCN,
the position is simplified because it is clear that the
physiological effects of the drug must be produced on
those enzyme systems, such as those involved in respiration, damage to which can produce an immediate result. But where there is a serious delay between administration of a drug and the emergence of its action, no
such simplification is possible and the whole enzyme

organisation of the cell must be suspect, including even those enzymes concerned in the mediation of genetic effects.

The case which we have been considering is the one in which we have the greatest knowledge of the organisation of enzymic processes. We cannot proceed very much further with the elucidation of the action of drugs in this manner until much more elaborate studies have been made on the enzymes concerned.

The Action of Various Enzyme Poisons on Different Physiological Processes

An alternative method of exploring the action of enzyme poisons is by considering their effect on different physiological processes. Table XVII shows the action of four different drugs on eight different physiological processes.

Of the four drugs, colchicine is the only one whose effect may be restricted to one physiological process only. The other drugs interfere with several physiological activities. As we have already noted, HCN, iodoacetate and mustard gas all have a strong action upon the metabolism of glucose, and theories have been put forward from time to time that their physiological effect is produced by interference with the metabolism of glucose. But when we consider the action of these substances on different physiological processes it is seen that their actions are not always the same. One of the most striking

TABLE XVII

THE ACTION OF FOUR ENZYME POISONS ON EIGHT DIFFERENT
CELLULAR ACTIVITIES

	Colchicine	M/1000 HCN	M/1000 iodoacetate	Mustard gas
Cellular respiration	—	Often 90% inhibition; sometimes none	Usually less than HCN	No action
Muscle contraction	—	None initially; lactic acid rigor later	Stopped almost immediately	No initial action; rigor later
Amoeboid movement	—	None initially	?	No initial action; stopped later
Ciliary movement	?	None initially	None initially	?
Cell division	Spindle formation inhibited	None initially; stopped later	Stopped	Spindle and chromosome abnormalities
Glycolysis	—	—	Stopped	Stopped eventually
Renal Secretion	—	Stopped	Stopped	No initial effect
Skin arterioles of frog	—	Dilated	Constricted	No initial effect

examples of this is in the action of cyanide, iodoacetate and mustard gas on the arterioles of the frog. Mustard gas produces practically no effect at all, HCN produces an almost immediate vasodilation and iodoacetate an almost immediate vasoconstriction. It is no doubt possible by the exercise of sufficient ingenuity, by taking account of the fact that these drugs act at different points upon the metabolism of glucose, to maintain that it is in fact interference with glucose metabolism which is the key point in the attack of all these substances upon cells. But this attitude smacks considerably of special pleading, and a great deal more analysis on the physiological as well as upon the biochemical level must be obtained before it can be accepted.

Classification of Drugs according to their Physiological Effect

The possibility exists that some useful principle might emerge from grouping drugs together which produce the same physiological change in cellular behaviour. But it is soon seen that a simple classification along these lines is not very enlightening. As examples we may take the inhibition of muscular contraction, the inhibition of mitosis, and the induction of lachrymation.

Of course, muscular contraction can in practice be inhibited by substances which act upon the neuromuscular junction. But as we are here concerned rather with cellular units, we shall not consider junctional inhibitors, but only the substances acting directly upon the muscle cell itself. As outstanding examples there are iodoacetate and potassium. Both of these substances rapidly give rise to a condition in which muscle will no longer contract. But they do so by quite different mechanisms, iodoacetate by preventing the synthesis of adenosine triphosphate, and potassium by reducing the excitability.

The formation of a properly functional spindle in mitosis can be prevented by a variety of substances. These include colchicine, urethan, arsenite, dithioglycerol, the β-chloroethylamines, and the sulphonamides. It seems extremely improbable that substances with such diverse chemical structures can be acting upon the same cellular mechanism. The action of the β-chloroethylamines, for instance, depends upon the loss of a chloride ion from the molecule with the consequent formation of a carbonium ion. There appears to be no analogous process possible with urethan. Then consider the two substances arsenite and dithioglycerol: the action of the former can actually be neutralised by a moderate amount of the latter.

Lachrymation is produced as a result of the action of substances upon the nerve endings in the conjunctiva. Substances producing this effect include soap, osmic acid and ethyl iodoacetate. It seems very improbable that all these substances can be acting upon the same enzyme system.

It therefore seems clear that not all the substances having a common physiological effect exercise that effect by an action upon a single enzyme system.

The Classification of Drugs in Terms of Enzyme Systems upon Which They Act

Whilst very little valuable information comes from classifying drugs according to the physiological effect they produce, since it soon becomes apparent that the substances producing a common physiological effect must do so by acting on a diversity of receptor systems, it still remains possible that substances acting specifically upon a certain enzyme system may produce a standard physiological effect. As examples we may take substances which act particularly upon SH enzymes, substances which are inhibitors of choline esterase, and substances which are inhibitors of hexokinase.

It seems likely, from the work of DIXON and his colleagues, that substances acting specifically on SH groups, such as those of triose phosphate dehydrogenase, are always lachrymators. In particular substances containing the groupings

$$R.CO.CH_2X, \text{ where } X = \text{halogen, and } R.CO.CH{=}CH_2$$

are often rather specific agents for SH groups. They include many of the substances which have been found useful in chemical warfare as lachrymators. On the other hand, not all lachrymators are capable of combining with SH groups: for example, soap does not.

Substances such as eserine act on choline esterase, and prevent the hydrolysis of acetyl choline, probably by competitive inhibition. As a result of this process marked

TABLE XVIII

THE ACTION OF VARIOUS SUBSTANCES, AS POISONS TO HEXO-KINASE, AND AS VESICANTS IN MAN

	Vesicancy	% Inhibition
$S(CH_2CH_2Cl)_2$	+ +	80
$OS(CH_2CH_2Cl)_2$	—	0
$S(CHCl.CH_3)_2$	—	0
$S(CH_2Cl)_2$	—	0
$S(CH_2.CH_2.CH_2Cl)_2$	—	0
$O_2S(CH=CH_2)_2$	+	60
$OS(CH=CH_2)_2$	—	0
$S(CH_2.CH_2OH)_2$	—	0
$S(CH_2.CH_3)(CH_2.CH_2Cl)$	+	45
$N(CH_2.CH_2Cl)_3$	+	70
$As\,Cl_2(CH=CH_2Cl)$	+ +	100
$As\,Cl_2(CH_2.CH_2Cl)$	—	0
$As\,Cl_2(CH_2.CH_3)$	+	45
CH_3Br	+ +	90
$CH_2{=}C.O.CO$ $\quad\mid\quad\mid$ $\quad CH{=}CH$	+	40

The inhibitory action was studied in the presence of $M/150$ glucose.

physiological effects ensue which include constriction of the pupil of the eye. Most of these substances act in a reversible manner, but studies by SAUNDERS, ADRIAN and DIXON have shown that the alkyl fluorophosphonates act upon choline esterase in a relatively irreversible manner and are also myotics. It thus seems likely that all substances which can produce a sufficient degree of inhibition of choline esterase have myotic activity.

Table XVIII is taken from the work of DIXON and NEEDHAM. It shows the action of vesicants on hexokinase in the presence of $M/150$ glucose. It will be seen that the substances having vesicant activity all poison hexokinase under these conditions, whereas those substances having no vesicant activity are not good poisons for hexokinase. The conclusion appears that all substances inhibiting hexokinase are likely to be vesicant.

From evidence of the type just given it seems likely that substances having a selective action upon a particular enzyme system are likely to display the same physiological activity.

The Mode of Action of Vesicants

As was noted in the previous section, the action of vesicants upon enzyme systems has been the subject of intensive study, both in England and in America. We shall examine these studies in a little more detail to show the great difficulty which is encountered in reaching a final conclusion as to the actual mode of action of any particular vesicant. We shall particularly consider lewisite and mustard gas, since these two substances have been the subject of the most consistent attack, both by enzymologists and by others. In the first instance we must distinguish between the local vesicant effect and the systemic effect of these substances. A survey of the literature shows that there is no common agreement between different workers as to the mode of action of these two substances. PETERS is of the opinion that the primary effect of mustard gas is upon the surfaces of skin cells, and possible upon the enzymes in those surfaces. The result of this action is the liberation of proteases which among other effects produce leucotaxin. On the other hand PETERS considers that the primary effect of lewisite is upon pyruvic oxidase.

DIXON and NEEDHAM, on the other hand, tend to the view that both substances exercise their primary effect, in common with other vesicants, upon hexokinase and perhaps other phosphokinases.

American workers in this field have reached somewhat different conclusions, in which they have been influenced by their observation that some proteases are as sensitive to these substances as are the phosphokinases. CORI and CANNAN appear to favour the view that the primary effect of mustard gas is upon the cell surface, and results in liberation of enzymes and changes in permeability. A recent American review concludes that "the specific chemical lesion is not yet defined by studies on inactivation of enzymes."

The effect of vesicants on the skin is so dramatic that

for long the systemic effects of mustard gas and of lewisite tended to escape study. Attention was particularly directed towards the systemic effects by CAMERON, who pointed out that in the long run the systemic effect might be much more important than the skin lesion. Subsequently many studies have been made of the basis of the systemic effects. PETERS inclines towards the view that the main effect is produced by the inhibition of pyruvic oxidase, whereas DIXON and NEEDHAM incline to the view that various phosphokinases may be concerned, amongst which may be part of the pyruvic oxidase system. Various cytological approaches have also been made to this problem. ROBSON, AUERBACH and KOLLER have indicated that the action of mustard gas may be primarily on the nucleus of a cell. They have shown that mustard gas is mutagenic and prevents mitosis. American workers have emphasised the action of mustard gas on cell membranes—red blood cells, Nitella, leucocytes, and lung cells have been studied and all appear to have their membrane properties significantly changed by amounts of mustard gas which would not exercise any profound effect upon intracellular enzymes. It has also been pointed out that there are many points of resemblance between mustard gas poisoning and radiation sickness.

From this survey of studies on vesicants it must be plain that whilst studies on enzyme systems are often very suggestive, yet it is often extremely difficult to obtain incontrovertible proof that a particular inhibition of one or more enzyme systems is in fact the result of a particular enzyme poison.

Biological Aspects of Enzyme Studies

When we look at the results of enzyme studies from the point of view of a biologist, we find a number of difficulties in applying these studies to cellular systems. These are:

1. Among the characteristic activities of many cells is the capability of concentrating foreign substances in vacuoles or granules. The biochemist is accustomed to adding a certain amount of enzyme poison to his enzyme solution or suspension, and then considering that the poison is uniformly distributed in the material under observation. But there is no such uniform distribution likely to occur in cellular systems, and the calculation of average concentrations for cellular systems is often fruitless, and even misleading. As a result of the concentration of drugs in particular localities in cells, a drug may never come into contact with the cellular systems which *in vitro* are the most sensitive to it.

2. Up to now the enzymes which have been mostly studied by biochemists have been cytoplasmic systems. It is much more difficult to make adequate studies on the enzymes of nuclei. Yet substances acting in very low concentrations are at least as likely to exercise

their effect upon nuclear as upon cytoplasmic systems.

3. In many cases it is possible to show a decline in the concentration of active enzyme in a particular tissue, subsequent to the administration of a drug. But usually this effect can only be demonstrated after the lapse of an interval from the time of administration of the drug. In such circumstances it is difficult to be sure whether the inactivation is produced by direct action of the drug upon the enzyme, or is an indirect effect ensuing from the action of the drug on some other enzyme system.

4. When the action of a given substance is studied on the enzyme systems of different organs, it is often found that the main effect appears to be selective for different enzyme systems in different organs.

5. In many cases some doubt as to an interpretation in terms of enzyme activity appears because a similar physiological effect can be produced by a mechanism which is primarily non-enzymic in its action. Thus vesication, which is produced by substances which are strong enzyme poisons, is also produced by friction, by heat and by cold.

6. If we attempt a straight-forward application of the results of enzyme studies, we should be tempted to say that drugs acting on the same enzyme system must produce the same effect. But this is seldom true. For example, vesicant substances all appear to poison hexokinase under certain conditions, and we might

expect the vesicles to be identical. But in fact on the cytological and histological levels, the vesicles produced by different substances appear very different. The general conclusion which we can reach at present is that drugs often exercise one or more of their modes of action through enzyme systems. Some drugs may exercise their action on enzymes only, or even on one enzyme only, but this degree of specificity is very rare.

REFERENCES

AUERBACH, C. and ROBSON, J. M., 1944: *Nature*, 154, 81.

BALDWIN, E., 1946: *Dynamic Biochemistry* (Cambridge University Press, London).

CLARK, A. J., 1937: *General Pharmacology* (Handbuch der Exp. Pharm. IV).

CULLUMBINE, H., 1947: *Nature*, 159, 151.

DIXON, M., 1948: *Nature*, 161, 226.

DIXON, M. and NEEDHAM, D. M., 1946: *Nature*, 158, 432.

DIXON, M., 1948: *Multi-Enzyme Systems* (Cambridge Univ. Press, London).

GILMAN, A. and PHILLIPS, E. S., 1946: *Science*, 103, 409.

KEILIN, D., 1929: *Proc. Roy. Soc.*, B. 104, 206.

KOLLER, P. C., 1947: *Symp. Soc. Exp. Biol.*, I, 270 (Cambridge University Press, London).

PETERS, R. A., 1947: *Nature*, 159, 149.

WORK, T. S. and WORK, E., 1948: *The Basis of Chemotherapy* (Oliver & Boyd, London).

CHAPTER V

The Actions of Narcotics

Introduction

In this chapter we shall deal with the action of narcotics on cells. This group of substances has not been studied as intensively by such a large number of chemists as have vesicants, but on the other hand it has been studied by people with much more diverse backgrounds. As a result there has been a very fruitful interplay of physical and chemical theories and a great variety of working hypotheses has been considered. Most of these hypotheses fall into three broad groups. The first of these involves an action on the cell surface or some other biologically important surface. The second emphasises the partition of narcotics between an aqueous phase and other phases of a lipoid character. The third considers narcotics as substances whose action is mediated primarily through enzyme systems.

Before proceeding in further detail, it will be as well to get some idea of what is referred to under the heading of narcosis. It is not a word with a single precise meaning. It includes such phenomena as the loss of consciousness, inhibition of a reflex, inhibition of the contractility of, say, heart muscle, inhibition of cell division, inhibition of ciliary movement, inhibition of respiration etc. Usually a given narcotic substance will produce all these

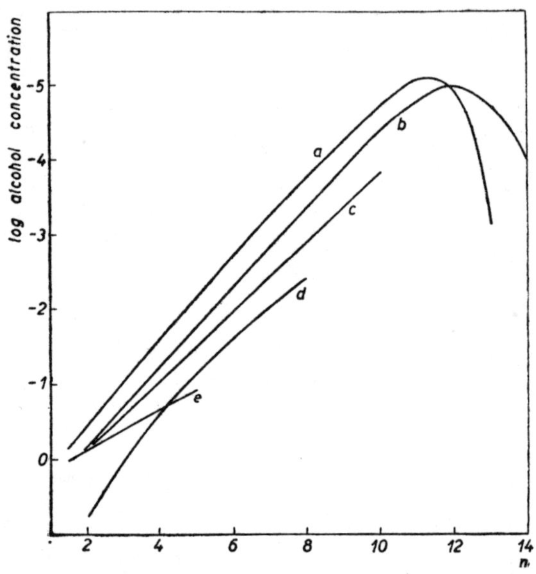

Fig. 18. Some narcotic actions of various primary alcohols. The concentration of alcohol producing a standard degree of narcosis is plotted against the number of carbon atoms in the alcohol. (a) Inhibition of swimming of tadpoles; (b) inhibition of hog ventricle; (c) concentrations reducing the surface tension of water by 10 dynes/cm; (d) concentrations lethal to *B. typhosus*; (e) immobilisation of *Paramoecium*

effects, but at different concentrations. Fig. 18 shows some characteristic results for the aliphatic primary alcohols. It will be seen that the different effects are produced by quite different concentrations of alcohol, and that the curves are not even all parallel to one another. From the facts already considered it is improbable that all the narcotic actions of a given substance are produced by the same mechanism. Hence it does not follow that what is established for a narcotic in one connection is necessarily involved in any other action involving the same narcotic.

Theories of Actions upon Surfaces

Permeability to narcotics. It is very common to find that the effectiveness of a narcotic increases as the oil-water partition coefficient is increased: the greater the relative concentration in oil, the greater is the narcotic action. The question naturally arises as to whether this is due to differences in cell permeability, for permeability usually also increases as the oil-water partition coefficient increases. This, however, is very unlikely to be the case, for most effective narcotics have a structure which allows them to penetrate quite rapidly into cells. For example, the aliphatic alcohols display an activity which usually increases by between 2.4 and 4.5 fold for each additional CH_2 group in the molecule. But all the aliphatic alcohols penetrate very rapidly into cells and the differences between the rates at which they permeate are far too small to allow for the increments in activity for each CH_2 group added to the molecule. It is probably quite seldom that the differences in narcotic activity in a homologous series can be attributed to the differences in rates at which the different members of the series penetrate into cells.

Is the action restricted to the cell surface? There is a good deal of evidence showing that substances, such as magnesium, cocaine and curare, act primarily on the cell surface, in some way modifying the excitability of the cell. Although the evidence as yet available is in no case conclusive, the best working hypothesis in the study of the action of these substances is probably that the action is restricted to the cell surface.

From time to time it has been suggested that the action of many other drugs is also restricted to the cell surface. Experiments which have received much attention are those of BRINLEY, HILLER and MARSLAND, who studied the action of narcotics and other substances by injecting them into the interior of *Amoebae*. They found that HCN, H_2S, picric acid and various of the common narcotics had no action when they were injected into the interior of *Amoebae*. This was true even if a narcotic was injected at a concentration which was sufficient to cause complete narcosis when an *Amoeba* was placed in the narcotic solution. An extreme example was that found with HCN. *Amoebae* placed in *M*/3000 HCN are killed in 24 hours, whereas the injection of *M*/100 HCN had no effect. The authors therefore argued that the action of these substances must be restricted to the external surface of the cell.

But there is a serious source of experimental error in work of the type just mentioned. All of the substances mentioned, including HCN, H_2S, picric acid and the narcotics, are of a type which pass through cell membranes very rapidly indeed. As a result, within a few seconds of making an injection into an *Amoeba*, practically the whole of the injected substance has diffused into the medium surrounding the *Amoeba*. Consequently experiments of this type are useless unless the substances injected can be relied upon to stay within the cell which is injected.

There is one set of experiments which is not invalidated by the diffusibility of the narcotic concerned. These were experiments made with paraffins as the narcotic substances. MARSLAND found that when a drop of olive oil is brought against the surface of *Amoeba dubia* it forms a cap on the surface. When an appropriate amount of a paraffin is dissolved in the olive oil, the *Amoeba* is fairly rapidly narcotised and ceases to move. But when droplets of this paraffin solution are injected into the interior of the *Amoeba*, no narcosis results. Hence in this case one may justly conclude that the narcotic effect of paraffins is exercised on the cell membrane.

Do narcotics change permeability to metabolites? A number of workers, particularly HÖBER, LILLIE and WINTERSTEIN, have concluded that narcotics exercise their action by decreasing the permeability of cells to essential metabolites. At the time at which the theory was put forward there was comparatively little experimental evidence available on the influence of narcotics on cell permeability. But a certain number of studies are now available. The more important of these are those of JACOBS and PARPART (1937), BÄRLUND (1938), and DAVSON (1940). BÄRLUND showed that ether tends to decrease cell permeability and that the decrease varies with the molecule which is considered. The effect is never very large. JACOBS and PARPART showed that butyl alcohol decreases permeability of the red cells of man, rat and rabbit to glycerol, but increases the permeability of the red cells of ox, sheep and dog to glycerol. DAVSON found that various narcotics increased the permeability of cat red cells to potassium and simultaneously decreased their permeability to sodium. These and various other results in the literature indicate that the effect on cell permeability of a given narcotic is sometimes to decrease the permeability, sometimes to increase it: the effect varies from cell type to cell type, species to species, and molecule to molecule. As the narcotic substances do not display any common effect on cell permeability it seems unlikely that the changes in permeability which they may induce are often prominent as effective modes of action in narcosis.

Is Traube's Theory of adsorption correct? TRAUBE noted that there is a close parallel between the effect of narcotic substances on surface tension and the physiological effect of the substances. As the number of CH_2 groups is increased, so surface tension is reduced and narcotic activity increases. The effect shows up in some cases rather clearly when the surface tension is studied of solutions of substances, all of which have roughly the same narcotic activity. There is a strong tendency for the solutions of equi-narcotic activity to have the same surface tension. Figure 19 shows an example of the parallel between

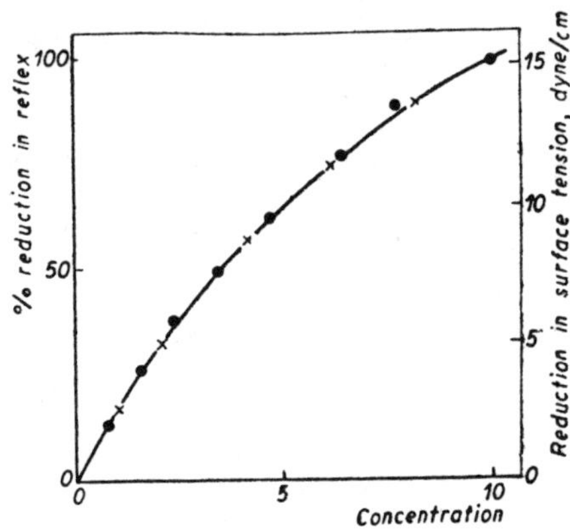

Fig. 19. The action of ethyl urethane in reducing the respiratory reflex in the cat (●) and in reducing the tension at the air-water interface (×)

reduction in the surface tension at the air-water interface and reduction in a reflex of the cat. TRAUBE therefore proposed that the narcotic substances exercise their effects primarily on surfaces, and that the effect is due to the reduction of surface tension. The strength of his case rested upon the fact that when a given substance is studied at different concentrations there is a parallel between narcotic activity and reduction in surface tension. And when different members of a homologous series are studied there is also a close parallel between the reduction of surface tension and the narcotic effect.

However there are three main criticisms which must be levied against this rather simple theory. These are:

1. The parallel between surface tension reduction and narcotic activity may be misleading. The reason for this is that in a homologous series the relative reduction in surface tension obtained with different members of the series is merely a reflection of the differences in the hydrocarbon moiety of the molecule. There are many other physico-chemical properties of molecules which also are a reflection of the hydrocarbon content of the molecule. For example, changes in the oil-water partition coefficient run closely parallel to changes in surface tension, for the simple reason that both properties closely follow the hydrocarbon content of the molecule. Thus from the studies of narcosis, one is justified in concluding that there is a close parallel between the narcotic effect and the amount of hydrocarbon in the molecule. But one is not justified in concluding that any particular individual physico-chemical property which runs parallel to the hydrocarbon content is necessarily the one through which the physiological effect is mediated.

2. Even if the effect of narcotic substances is exercised at an interface, studies on the air-water interface may very well be irrelevant. The important interfaces of a cell are oil-water, protein-water, nucleic acid-water,

etc. Consequently the parallel between air-water surface tension and narcotic effect noted by TRAUBE is quite likely to be fortuitous.

3. As MEYER has shown, some narcotics have no action on the air-water surface tension, for example methane and nitrous oxide are both without action upon the air-water surface tension, although they can exercise a potent narcotic effect. It is, therefore, clear that whilst some narcotics may exercise their effects upon an interface having properties similar to the air-water interface, other narcotics must act in a quite different manner.

Consequently one can conclude that a simple theory of absorption such as that of TRAUBE is unlikely to be of major importance, except in peculiar cases. This however, must not blind us to the fact that adsorption at interfaces may indeed be a very important part of the action of many narcotics. But if so the mechanism is not so simple as TRAUBE has suggested.

Theories based on oil-water partition effects

A number of workers have put forward the hypothesis that many narcotic substances exercise their physiological effect by virtue of changes which occur in the organisation of essential structures of the cell as a result of the dissolving of the narcotics in oily phases which are part of these structures. For example, suppose we consider the substances

$$CHCl_3$$

$$CH_3.CH_2OH$$

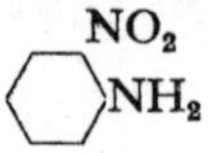

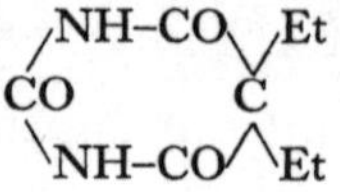

First inspection of these formulae suggests that it would be impossible for these substances to act through the same chemical mechanism, but if we make the hypothesis that these substances act after dissolving in a lipoid phase, then the first thing we must do is calculate what are the relative concentrations of these substances in the lipoid for solutions having the same narcotic activity. It is surprising to find that although the structures of these substances are quite different, and although their equinarcotic concentrations in water are quite different, the concentration of these substances in lipoid is often practically the same.

Tables XIX and XX show two examples of calculations made by K.H.MEYER (1937), in which he develops the OVERTON-MEYER hypothesis in a fairly precise manner. One column in these tables shows concentration of material which produces a given degree of narcosis. The other column shows the concentrations of the different substances in an oily phase which would be in equilibrium with the narcotic concentration in air or water, as the case may be in the two tables. For the results shown in Table XIX where the narcotic is present in air, the concentrations of different narcotics producing a standard effect show a variation of 740 fold, whereas the corresponding equilibrium concentrations in oil show a variation of only 1.8 fold. Similarly in Table xx the concentrations of the different narcotic substances in water which produce a given degree of narcosis show a 7000 fold variation. But the corresponding equilibrium concentrations in oil show a variation of only 2.5 fold. It is

TABLE XIX

THE CONCENTRATIONS IN AIR OF VARIOUS NARCOTICS REQUIRED TO PRODUCE A GIVEN DEGREE OF NARCOSIS IN MICE, AND THE CORRESPONDING EQUILIBRIUM CONCENTRATIONS OF THE NARCOTICS IN OLIVE-OIL

	Concentration in air; vols. per cent	Corresponding concentration in olive oil; mol/litre $\times 10^2$
Methane	370	8
Nitrous oxide	100	6
Acetylene	56	5
Ethyl chloride	5	7
Ether	3.4	9
Methylal	2.8	8
Carbon disulphide	1.1	7
Carbon tetrachloride	0.6	7
Chloroform	0.5	9
Range of variation	740 fold	1.8 fold

thus clear that there is a very close correspondence between the production of a given degree of narcosis, and the production of a standard concentration of molecules in a lipoid phase. The detailed chemical structure of the molecules does not appear to be of any importance. It is the features of the molecules which determine their oil-water partition coefficient which also determine narcosis.

The partition of a substance between oil and water can

TABLE XX

THE CONCENTRATIONS IN WATER OF VARIOUS SUBSTANCES REQUIRED TO PRODUCE A GIVEN DEGREE OF NARCOSIS IN TADPOLES, AND THE CORRESPONDING EQUILIBRIUM CONCENTRATIONS OF THE SUBSTANCES IN OIL

	Concentration in water; mols per litre	Corresponding concentration in oil; mols per litre $\times 10^2$
Ethyl alcohol	0.33	3.3
Propyl alcohol	0.11	3.8
Butyl alcohol	0.03	2.0
Valeramide	0.07	2.1
Antipyrin	0.07	2.1
Pyramidon	0.03	3.9
Ether	0.024	5.0
Benzamide	0.013	3.3
Salicylamide	0.0033	2.1
Luminal	0.008	4.8
o-Nitraniline	0.0025	3.5
Carbon disulphide	0.0005	3.0
Chloroform	0.00008	2.6
Thymol	0.000047	4.5
Range of variation	7000 fold	2.5 fold

be regarded as determined by the sum of the differences in free energy of its component groups in the two media. From the data just given it is evident 1. that there is a parallel between equi-narcotic action and concentration in a lipoid phase which is not upset by variation in either polar or non-polar groups, and 2. that no structural specificity in the molecules is involved. It, therefore, seems very improbable that adsorption at an interface is the mode of action in the two instances of narcosis which have just been given. Equally, it seems highly probable that narcosis is produced in these two instances by the production of a critical concentration of foreign molecules in a bulk phase consisting mainly of hydrocarbon.

At this point, we must ask the question, where is the lipoid phase in the cell? It may perhaps be the interior of the cell membrane, the lipoid micelles, or possibly a conglomeration of the hydrocarbon residues of protein. Possibly all three sites may be involved in different instances of narcosis. And no doubt there are some sites which have not yet been thought of.

It is important to note that this theory is not incompatible with the involvement of enzymes in narcotic action. The action upon any of the three sites just mentioned may well produce an ultimate action upon the degree of activity of an enzyme system.

Theories Based on Actions on Enzymes

We have previously noted that urethane acts upon dehydrogenases and will inhibit respiration for this reason.

But the linkage between respiration and other cellular effects is complex, as is well illustrated by the fact the

the same physiological effect can sometimes be achieved either by increasing or by decreasing the respiration rate! For example, dichlorophenol and urethan both stop cell division at a certain minimal concentration, which is of course different for the two substances. But

TABLE XXI

THE NARCOTIC ACTION ON GUINEA PIGS, AND THE INHIBITION OF OXYGEN UPTAKE BY GUINEA PIG BRAIN, OF CERTAIN BARBITURATES

	Narcotic action	% Inhibition of O_2 uptake by brain
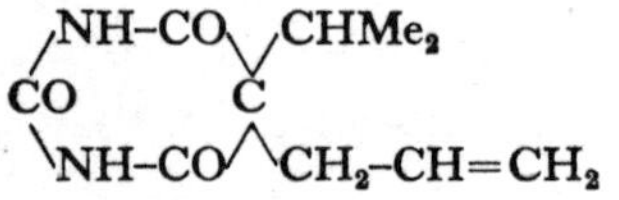 NH–CO, CHMe₂ / CO — C / NH–CO, CH₂–CH=CH₂	+ +	40
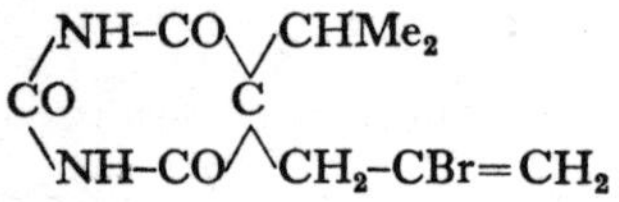 NH–CO, CHMe₂ / CO — C / NH–CO, CH₂–CBr=CH₂	+ +	50
NH–CO, CHMe₂ / CO — C / NH–CO, CH₂–CHBr–CH₃	negligible	o

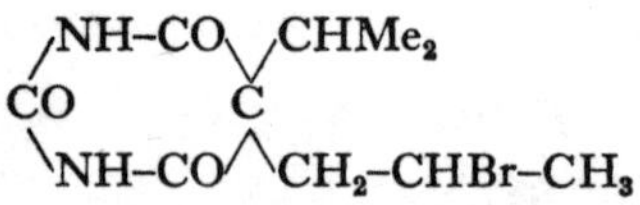

dichlorophenol at the concentration at which it stops cell division increases the respiration of *Arbacia* eggs by more than 100%, whereas urethan at the concentration at which it inhibits the division, reduces the respiration by about 75%!

Amongst the more striking studies of the effects of

narcotics upon respiration is that of QUASTEL upon the respiration of the brain tissue. Table XXI shows some of his results. Inspection of this Table indicates that with the substances he was investigating a difference of one double bond in structure was sufficient to cause profound modification of narcotic activity, and an equally profound modification in the ability to inhibit oxygen uptake by brain tissue. It is clear that the very small difference in partition coefficient produced by a difference in structure of one double bond could not possibly account for the differences of activity of these different molecules. One therefore wonders whether this is an instance in which a specific structure of the molecule producing narcosis is important, and whether this structure is not specific for one of the enzyme systems involved in respiration.

Some doubt however is thrown upon this by the results given in Table XXII. This shows in one column the concentration of narcotic which is necessary to produce a given degree of narcosis in the rat, and in the other column, the degree to which the uptake of oxygen by brain tissue is inhibited by the narcotic concentration. If all the narcotics are acting in the same manner, one would except them to inhibit the respiration of brain by the same amount when present in equi-narcotic concentrations. But, in fact, urethan produces only one-fifth of the inhibition produced by an equi -narcotic concentration of avertin. The evidence, therefore, is not very convincingly in favour of the idea that narcotics are acting upon a single site when acting upon brain tissue. It is possible that the results would have conformed more precisely to theory had the brain tissue been in a normal condition, instead of being in slices. But, never-

TABLE XXII

THE CONCENTRATIONS OF NARCOTICS PRODUCING A GIVEN DEGREE OF NARCOSIS IN THE RAT (I.E. EQUINARCOTIC CONCENTRATIONS), AND THE PERCENTAGE INHIBITION OF THE RESPIRATION OF RAT BRAIN SLICES PRODUCED BY THESE CONCENTRATIONS OF NARCOTIC

	Equi-narcotic concentration in rat	% Inhibition of brain respiration
Ethyl urethan	0.022	6
Chloral hydrate	0.0013	10
Luminal	0.00079	15
Chloretone	0.0010	20
Avertin	0.00106	31

theless, one cannot help being highly suspicious of a simple theory of action on one site.

Further investigations by QUASTEL have shown that the effects of narcotics on respiration are probably produced by their effect on carbohydrate metabolism, which is of exceptional importance in the respiration of brain. But he found that in the low concentrations producing narcosis in mammals, there is no significant direct effect

of the narcotics on the dehydrogenases themselves.
The effect of the narcosis is on an earlier step as indicated by the following diagram:

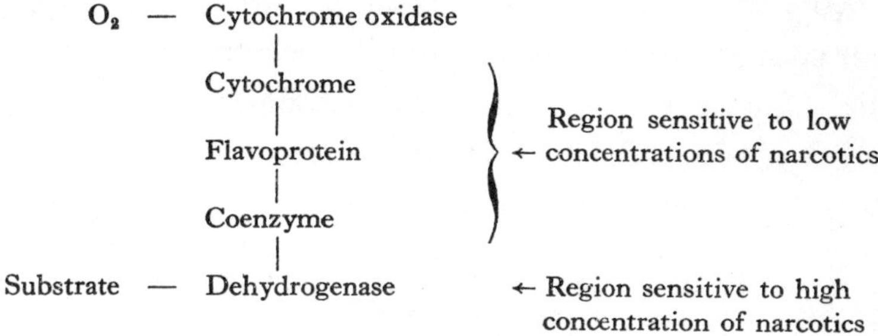

The views which QUASTEL has developed, outlined above, refer only to the action of narcotics upon the central nervous system of mammals, particularly of rodents.

JOHNSON and his colleagues have made some very stimulating studies from a quite different point of view on the effect of narcotics on bacteria. They found that narcotic action can be antagonised by high pressure. It had been noted that when protein denatures there is an increase in volume. If this is prevented by compression, no denaturation occurs. JOHNSON, therefore, argues that since the action of narcotics can be reversed by high pressure, the narcotics act by denaturing proteins and not by adsorption on specific active centres of enzymes. This view is, of course, quite compatible with the oil-water partition coefficient hypothesis of OVERTON and MEYER, for it is quite clear that lipoid substances can denature proteins. They probably do this mainly by changing the organisation of the lipoid residues in the polyptide chains which make up the proteins. This theory of JOHNSON's has the advantage that it also allows for species variation, since when a given enzyme is taken from different organisms, it usually shows marked variation in ease of denaturation, according to the source from which it has been taken.

REFERENCES

BARLUND, H., 1938: *Protoplasma*, 30, 70.

BRINLEY, F. J., 1928: *J. Gen. Physiol.*, 12, 201.

CLARK, A. J., 1937: *General Pharmacology* (Handbuch der Exp. Pharm. IV).

DAVSON, H., 1940: *J. Cell. Comp. Physiol.*, 15, 317.

DAVSON, H. and DANIELLI, J. F., 1943: *The Permeability of Natural Membranes* (Cambridge University Press, London).

HILLER, S., 1927: *Proc. Soc. Exp. Biol. and Med.*, 25, 305.

HOBER, R., 1945: *Physical Chemistry of Cells and Tissues* (Churchill, London).

JOHNSON, F. H., BROWN, D. E. and MARSLAND, D., 1942: *J. Cell. Comp. Physiol.*, 20, 269.

LILLIE, R. S., 1923: *Protoplasmic Action and Nervous Action* (Chicago).

MARSLAND, D., 1934: *J. Cell. Comp. Physiol.*, 4, 9.

MEYER, K. H., 1937: *Trans. Faraday Soc.*, 32, 1062.

QUASTEL, J. H., 1943: *Trans. Faraday Soc.*, 39, 348.

TRAUBE, I., 1935: *Biochem. Z.*, 277, 39; 282, 444.

TRAUBE, I., 1937: *Trans. Faraday Soc.*, 32, 1066.

WINTERSTEIN, H., 1926: *Die Narkose* (Berlin).

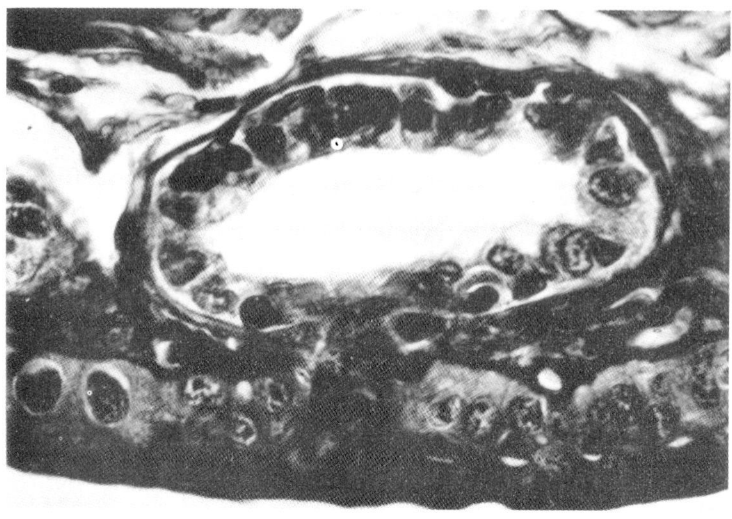

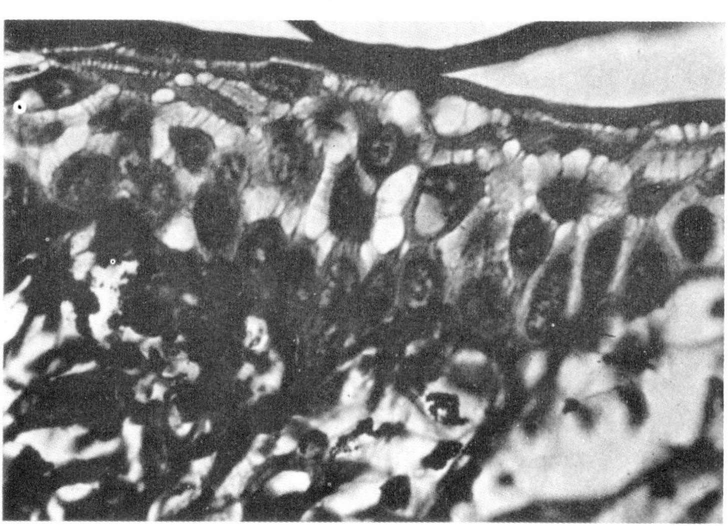

Plate II. Vesicles in the skin of the frog, produced by different agents. *a* is a normal skin. In *b* the prickle cells adhere firmly to oneanother, and to the dermis; the cornified layer splits away from the prickle cells.

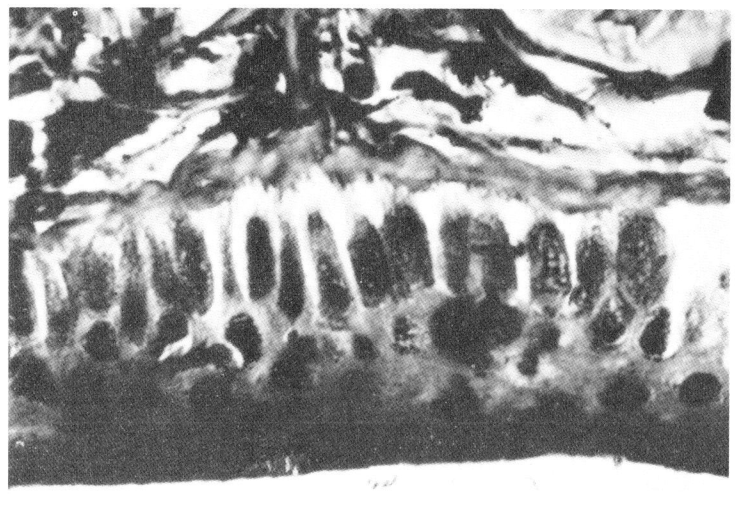

c

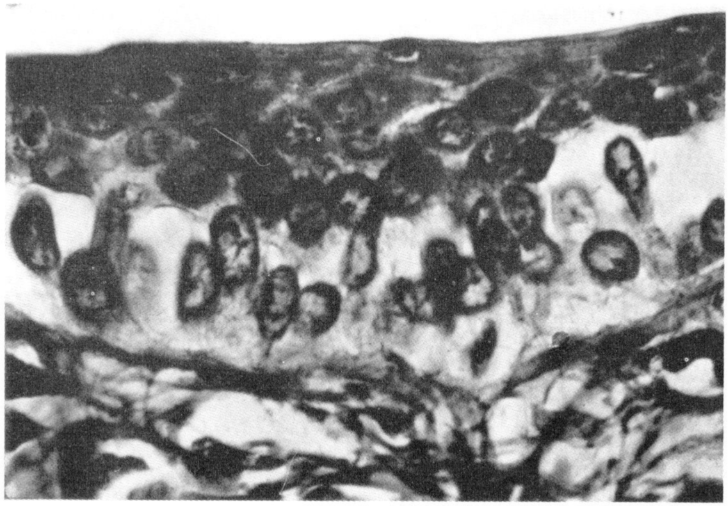

d

In *c* the whole epidermis is splitting away from the dermis, and in *d* the split is occuring in the middle of the prickle cell layer. The three types of vesicle *b*, *c* and *d* are characteristic of the different agents producing them (see p. 120 and 121).

CHAPTER VI

Responses of Cells on the Biological Level

Introduction

It can be concluded from the previous chapters that if the attempt is to be made to develop new drugs on a rational basis, rather than on the hit or miss principle, two types of research unit are required. One of these types is necessary to study the general physico-chemical properties of the cells of mammals and of parasites, in particular such properties as permeability, secretion mechanisms, excitability phenomena and adsorption effects. The other type is more biochemical in character, and is particularly required for the study of enzyme effects. But this type of research in only a part of that which can be deduced to be necessary from the cytological point of view. There is also a great need for more frankly biological studies of cells. Altogether too little is known of what may, by analogy, be called the natural history and ecology of cells, and their responses on the biological level.

The Nature of Biological Responses

The biological responses of cells to drugs are occasionally highly specific and can only be elicited by a very small

Cell Physiology 8

range of compounds. But more often the response may be evoked by a wide variety of substances and indeed by a wide variety of types of stimuli. For example, the characteristic response of nerve or muscle may be elicited by heat, cold, pressure, cutting, potassium, various organic bases and by electrical shocks. Similarly, the complex series of cellular events resulting in vesication may be elicited by heat, cold, friction, bacterial toxins, arsenoxides and β-chloroethylamines. It is very tempting to conclude that all the agents producing a given effect must be acting upon the same mechanism, and that by contemplation of what system would have the ability to respond to all of these agents it will be possible to deduce the nature of the reactive mechanism in the cell. However, more careful analysis of the biological facts frequently indicates that the different agents acting upon a cell may be reacting with quite different systems, although there may be a great deal in common in the final results of such action. For example, a muscle may be stimulated to contract both by treatment with potassium chloride and by treatment with acetyl choline: but the quantities of these substances which are required to produce contraction are so remarkably different that it is impossible to credit that they act upon the same system. Then again, when the details of vesication are considered, one finds that there are marked differences between the vesicles produced by different agents. Plate II (pages 116 and 117) shows three types of vesicle, produced by three different types of agent acting upon the skin of the frog. In the one case, the skin has split between the cornified layer and the prickle cells; in the second case, the skin has split in the middle of the prickle cell layer; and in the third case, the prickle cell layer and the cornified layer have become detached from the dermis. If the action of these three agents had been assessed as either "vesication" or "not vesication", all three would have been assessed as vesicants and presumed to act through the same mechanism. But the details of their biological action are in fact so different that it must be concluded that only part of their action at most can be exerted upon a common mechanism.

From these examples, as from many others which could be adduced, it must be concluded that a cell or tissue is commonly designed to fulfil a particular purpose, and that it responds to stimuli of very diverse types in a manner which is characteristic of the cells involved, and not necessarily of the nature of the stimulus. Different stimulating agents may act upon quite different cellular systems, but the design of a cell is commonly such that these mechanisms are funnelled to result in the elicitation of a response characteristic of the particular design of cell. With this point in mind we can proceed to examine a number of types of biological response in rather more detail. These will include artificial parthenogenesis and mitotic abnormalities, and the responses of genetic systems to drugs.

Artificial Parthenogenesis

In Fig. 20 we see arranged a list of agents which are competent to cause artificial parthenogenesis. The first group including substances such as fatty acids, saponin and organic bases, have in common the possibility that they may act upon the surface of the cell causing, as J. LOEB

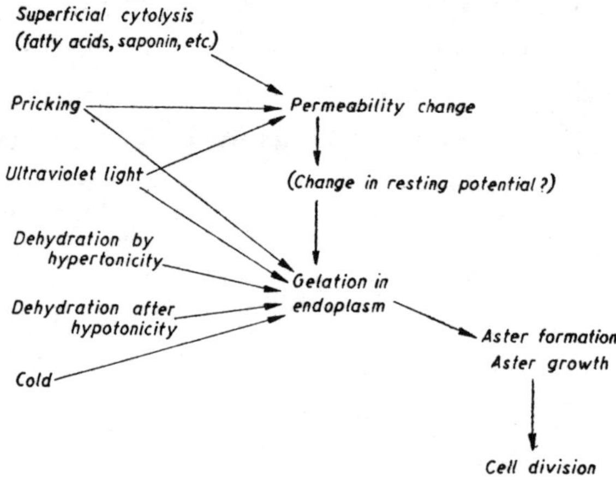

Fig. 20. Mode of action of agents causing artificial parthenogenesis

suggested, a superficial cytolysis to arise, or to become incipient. With some species activation may also be secured by totally different methods such as pricking, the use of ultra-violet light, treatment with hypertonic or hypotonic solutions, and treatment at low temperatures. So far as is known, there is no common mechanism which can be acted upon by these diverse agents and processes. But all the reagents ultimately lead to cell division, and it seems probable that they can do so in the absence of the cell nucleus. The characteristic required for cell division appears to be a process involving an activity of the cortical gel layer of the egg and also the formation of two asters in the cytoplasm. The reagents also have in common the fact that if used somewhat excessively they cause not two asters but many asters to appear in the cytoplasm. It seems probable, therefore, that the various agents act at different stages in a chain of processes which leads up to gelation of the endoplasm and a concomitant activity of the cortical gel. Beyond this it is not possible to proceed at present. If the mechanism of parthenogenesis is to be understood it will probably be necessary to work backwards from the biological response of aster formation to the details of the mechanism of activation involved in each particular type of reagent.

Mitotic Poisons

A considerable number of drugs have been classed as mitotic poisons, and in the chemotherapy of cancer increasing attention is being paid to the action of drugs on mitosis. Mitosis is a complex process and can be regarded as consisting normally of at least eight steps. These are:

1. Division or duplication of chromosomes (involving duplication of each gene).
2. Spiralisation of the chromosome accompanied by con-

densation upon it of nucleic acid, and a fading away of nucleoli.

3. Division of the centrosome and breakdown of the nuclear membrane.
4. Formation of the spindle and of the equatorial plate.
5. Division of the equatorial plate and movement of chromosomes toward the centrosomes.
6. Division of the cytoplasm.
7. Reformation of nuclear membranes.
8. Despiralisation and loss of nucleic acid from the chromosomes, and reconstitution of nucleoli.

Obviously mitosis is a very complex process, and if mitotic poisons act directly it is obvious that derangement will occur at many points, and through interference with many quite distinct phenomena. It is, of course, quite possible that some, if not all, of the known mitotic poisons exercise their effect, not by acting upon the specific processes concerned exclusively in mitosis, but on one or more of the processes which, amongst other things, supply the necessary energy for the different stages of mitosis.

Among the commoner effects of mitotic poisons are:

1. The adhesion of chromosomes to one another (commonly called "stickiness") and the failure of daughter chromosomes to separate completely from one another. Such phenomena commonly lead to
2. The breaking of chromosomes, and the formation of chromatin fragments which may not become at-

tached to the spindle at subsequent divisions. These phenomena lead to unequal distribution of chromatin between daughter cells.

3. Failure of spindle formation. This may lead to polyploidy, and on occasion, if the chromosomes have failed to aggregate, to the formation of micronuclei, each organised by one or a small group of chromosomes.

4. The formation of multi-polar spindles, perhaps due to multiple division of the centrosome, or perhaps to the formation of new centrosomes. The usual result of this process is the uneven partition of chromatin between daughter cells, or occasionally the formation of a multinucleate cell.

5. Chromosomes adhesions may prevent the complete separation of the nuclei, and the result may be that cell division does not go to completion.

6. All the processes through which the spindle goes may be slowed down.

7. Over-spiralisation, and failure to despiralise after cell division, may occur.

8. As a result of fragmentation and breaking, translocation may occur.

9. Cell division may fail to go to completion, resulting in the formation of a bi- or multinucleate cell.

In view of the complexity of the phenomena involved in mitosis and the great variety of abnormalities which may occur, it is obviously insufficient to classify drugs as mitotic poisons because they have the common effect of deranging of mitosis. At present our knowledge of the biochemical, biophysical and biological characteristics of the various known mitotic poisons is quite insufficient for an analysis of their actions, but some hints of the type of sub-groupings that may be expected are available. For example, colchicine differs from most of the other mitotic poisons in that, in low concentrations, its action appears to be almost entirely restricted to suppression of spindle-formation. It is therefore probable that it attacks the cell in a different way from, say, some of the nitrogen mustards, with which bridge formation and fragmentation are very prominent phenomena.

ÖSTERGREN has shown that a very wide variety of organic compounds may cause stickiness (adhesion of chromosomes) and suppression of spindle-formation. The equitoxic concentrations of these substances are roughly proportional to their oil-water partition coefficients, and it may be that in such cases the toxic effect is produced by a mechanism similar to that discussed in the chapter on narcosis. A small number of substances are active as mitotic poisons in concentrations much smaller than would be suggested by their oil-water partition coefficients. This suggests that their action is a specific action on the cell at some point more intimately related to mitosis than is the generalised effect of organic compounds which is proportional to the oil-water partition coefficients. It does not, however, necessarily

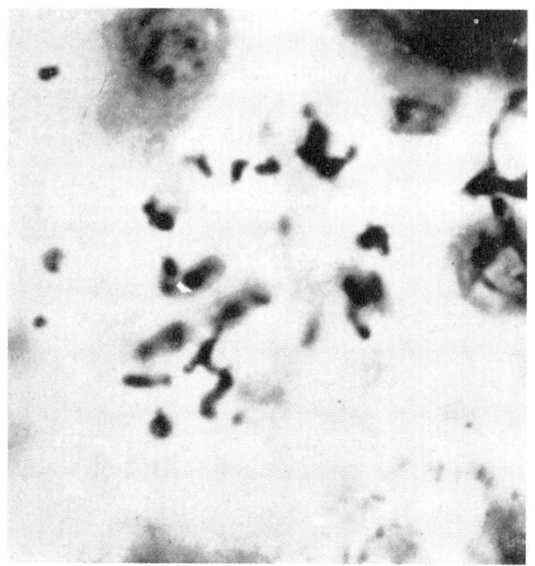

a

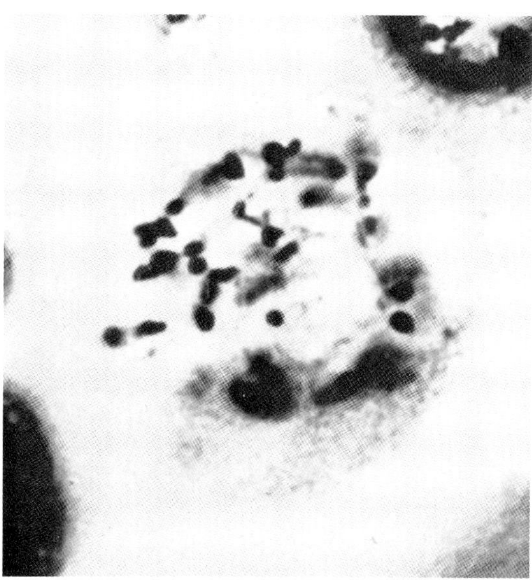

b

Plate III. A comparison of the distributions of deoxypentose nucleic acid and alkaline phosphatase in cells of the rat Walker sarcoma, poisoned with a nitrogen mustard. *a* Failure of spindle formation –

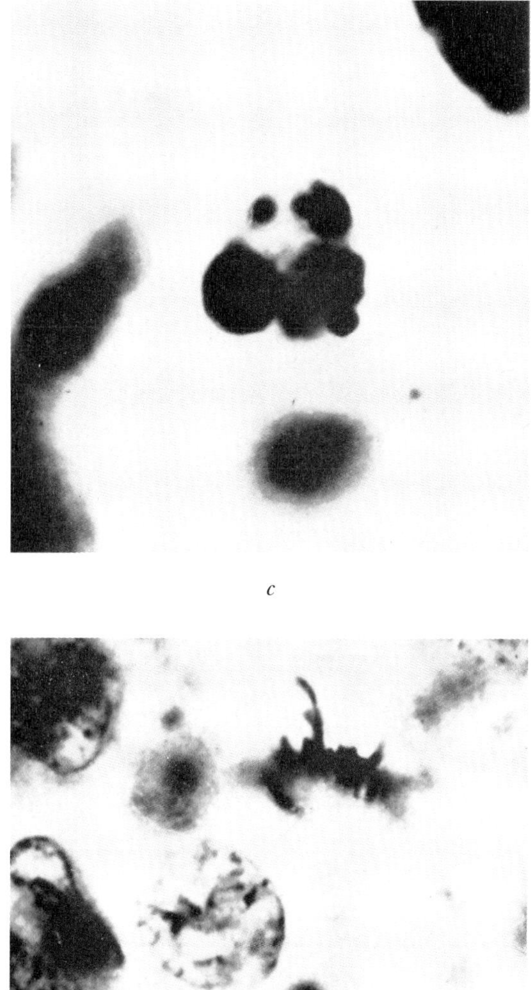

Feulgen. *b* Failure of spindle formation – phosphatase. *c* Pycnotic degeneration – alkaline phosphatase. *d* Pycnotic degeneration – Feulgen. (see p. 134).

follow that the specificity represents an exceptional potency for reaction with, say, a chromosome or spindle constituent. The apparent specificity may in fact be due to the operation of a process not even distantly related to mitosis. Thus, those compounds which readily fit into the secretory patterns of cells, and so are found in disproportionately high concentrations inside cells by comparison with other apparently similar compounds, would be expected to display an apparently abnormally high toxicity if these high concentrations of drug are effective in those regions of the cell which are intimately connected with mitosis.

Among the substances having a disproportionately high activity by comparison with their oil-water partition coefficients are urethan and some members of the nitrogen mustard series. It has been suggested that urethan acts by interfering with nucleic acid metabolism, and that the nitrogen mustards may also act in this way. The evidence in support of this hypothesis is at present very slender.

An interesting example of the difficulties which may be encountered is found in the action of arsenical compounds on mitosis. Substances such as sodium arsenite and phenyl arsenoxide are fairly strong mitotic poisons, and share this property with iodoacetamide. One of the few properties which these three substances have in common is that of combining vigorously with SH groups, and it has therefore been suggested by RAPKINE that SH groups are of particular importance in mitosis. It is not clear whether the SH groups concerned are to be regarded as part of the chromosomes, part of the spindle proteins or part of the general enzyme systems of the cell. In support of the contention that SH groups are particularly involved, it has been found that B.A.L. (dithioglycerol) is able to reverse the toxic effects produced by sodium arsenite and phenyl arsenoxide, but not the toxic effect of iodoacetamide. The position, however, is more than a little complicated by the fact that B.A.L. is itself a mitotic poison, and its toxic action can be reversed with arsenoxide.

The present position is that whilst it is clear that substances having a fairly high degree of specificity for SH groups are mitotic poisons, it is not yet clear that the true point of attack of these substances is in fact upon SH groups. And beyond this, it is still less clear whether the supposed SH groups are, as some contend, part of the spindle proteins, or whether they may not even be so distantly related to the actual processes of mitosis as to be the SH groups of enzyme systems concerned in the mobilisation of energy, such as the SH groups of phosphokinases.

Substances which are generally regarded as having an action upon respiratory mechanisms involved in the mobilisation of energy, quite commonly have a very profound effect upon mitosis. This is true of HCN, of phenols such as hydroquinone, of urethan and of the quinones. Whilst certain members of these groups, such as urethan, may on some cells act in such low concentration as to suggest that an action upon non-respiratory mechanisms is involved, others, such as phenyl urethan, act at a level of concentration which appears to be similar to that involved in the inhibition of respiration.

The β-chloroethylamines. It is now well-known that many compounds containing two β-chloroethyl groups are potent mitotic poisons. The first of these to be examined was mustard gas itself. This was shown by KOLLER, in studies on *Tradescantia*, to produce chromosome breaks, failure of division, the lagging of chromosomes and the formation of bridges between the separating groups of daughter chromosomes in anaphase and telephase. HUGHES and FELL have made a particular study of these phenomena in tissue culture, and show that spindle abnormalities, such as tripolar spindles, are very common.

More recently attention has been focussed upon the use of the so-called nitrogen mustard compounds of the general type $R.N:(CH_2.CH_2Cl)_2$. These compounds in some instances have a relatively selective effect upon the growth of some types of tumours, and as KOLLER has shown, this effect is probably largely produced by the action of these compounds upon mitosis. Evidence has recently been obtained by REVELL that the hetero-chromatic regions of the resting nucleus are particularly susceptible to attack. On the chemical level, as with the sulphur mustard compounds, activity seems to be associated with the ability of the chloride ion to dissociate from the β-chloroethyl group, leaving a positively charged carbonium ion which has the capacity to react with many cell components. Although many speculations have been put forward about the biochemical mechanism through which these compounds act, the evidence so far available is too sparse to enable any definite conclusion to be drawn.

It is possible that a useful guide to the mode of action of many mitotic poisons, including the nitrogen mustards, will be obtained from cytochemical studies. At present such studies are very few, and so far have been more indicative of the degree of involvement of chemical processes in the biological response itself than of the biochemical mechanism initiating the biological response. For example, some comparative studies have been made of the FEULGEN reaction for deoxypentose nucleic acid and of the reaction of TAKAMATSU and GOMORI for alkaline phosphatase. Some of these results are shown in Plate III (pages 128 and 129). It will be seen that the anomalies produced in the distribution of nucleic acid are strikingly similar to those produced in the distribution of phosphatase.

It must, of course, be remembered that the identification of the biochemical mode of action of a compound such as nitrogen mustard may prove to be extraordinarily difficult by the biochemical techniques which are yet available. On theoretical grounds one would be inclined to predict that most of the mitotic abnormalities which are seen as a result of the action of a dose of nitrogen mustard could originate as the result of the failure of one or a few genes either to reproduce, or to function. The reproduction, or functioning, of a single gene can probably be inhibited by combination with one molecule of nitrogen mustard. Particularly significant in this connection is some of the recent work of HERRIOT on viruses, the reproduction of which he shows to be much more sensitive to mustard than is any other biological process so far examined. If it should prove to be true that the most important site of action is upon the genes and that the action is exerted by a small number of molecules of nitrogen mustard, then quite exceptional methods will be necessary to detect the exact site of action. Furthermore, the identification of the chemical action exerted upon the gene will be very difficult to ascertain, partly because of the quantities involved, and partly because the reactions between the gene and the drug need not necessarily be those which are regarded as in the main course of chemical reactivity of the drugs concerned. Where a few molecules only out of a relatively large dosage can have a definitive action upon a biological process, it is perfectly possible for the key reaction involved to be amongst those which are normally classified by the organic chemist as "side reactions", and be usually even less understood than the characteristic reactions.

Reproduction of Bacteria and Viruses

A number of substances which have a very potent effect upon the multiplication of some types of bacteria and plants have also the property of producing at least some of the phenomena which have been mentioned as characteristic of the substances which have been classified as mitotic poisons. Amongst the bacteriostatic substances are, for example, the sulphonamides, which have been shown to prevent cell division and cause polyploidy in onion root tips. It may very well be that examination of such cases would assist considerably in understanding the mechanisms which may be involved in mitotic poisoning. But the analysis of these cases is not likely to be particularly simple. Thus, according to the views developed by WOODS and FILDES, the sulphonamides usually exercise their bacteriostatic effect by preventing the utilisation of para-aminobenzoic acid: and GALE has recently suggested that the primary action of penicillin is to prevent the uptake of glutamic acid. The question therefore arises as to whether the action of such substances as mitotic poisons involves the same inhibitions as are supposed to be concerned in bacteriostasis, or whether some other quite different actions are involved when they act as mitotic poisons.

It is, of course, possible that the bacteriostatic action is, in fact, identical with the action upon mitosis. But it cannot at present be assumed that the phenomena of multiplication of genetically active units in bacteria involve all the steps concerned in mitosis, particularly when these events are viewed from the morphological level. When we turn to the multiplication of bacteriophages and viruses it is possible that we are dealing with a different or at all events much simplified process, with which the possible routes of interference are more restricted. It is quite likely that the reason why the common bacteriostatic agents and mitotic poisons seem to be of little use in the treatment of virus diseases is that some of the specialised processes involved in mitosis simply do not occur in the reproduction of viruses, and that it is these rather specialised processes which are primarily attacked by the known poisons.

If it is indeed the case that virus reproduction is a simpler process than mitosis, it may be necessary to look for rather different physico-chemical phenomena as possible modes of attack in the designing of substances which will prevent the production of viruses. For example, the study of viruses *in vitro* may well lead to the discovery of types of substances which are selectively adsorbed upon them: such substances may well prevent the reproduction of viruses. Then certain viruses have been shown to attach themselves to particular points on the surfaces of cells, and it has also been shown that particular substances may be necessary to secure this attachment. From this, two possibilities arise for the design of competitive substances: one type of substance which would be of interest would be those compounds which can adhere more vigorously than does the virus to the regions of the cell surface which are specifically virus-adsorbing; then, if a substance like tryptophane is required for the adhesion of a virus to a cell, a satisfactory inhibition of virus activity might be obtained by using a competitive substance such as methyl tryptophan.

The relationship of mitotic poisons to mitotic stimulators is also one which would probably repay more detailed examination. Whilst it is possible that some stimulating substances may be rather highly specific, other substances which act as mitotic stimulators have a relatively generalised effect. Xanthopterin has an effect which is mainly restricted to the kidney and a smaller effect upon the bone marrow. But oestrone has a relatively generalised effect, producing an increase in mitotic rate in practically all cell lines capable of mitosis. Despite this, oestrogens act as mitotic poisons for the cells involved in cancer of the prostate and post menopausal breast tumours. One wonders whether the mechanisms involved in mitotic poisoning and mitotic stimulation are related or not. It is, of course, no new thing to find that a substance may apparently have diametrically opposite physiological effects under different circumstances: for example, adrenaline is vaso-constrictor for some arterioles and vaso-dilator for other arterioles.

Nuclear and Cytoplasmic Drug Action

It is not unreasonable to suppose that some substances will exercise their main action upon the cytoplasm of cells and others will have their main action on the nuclei. We can form an initial impression of what the differences may be between these two lines of approach by considering studies upon cells from which the nuclei have been removed. From such experiments we know that cells without nuclei may retain their form, conduct electrical impulses, exhibit the phenomena of amoeboid movement, phagocytosis and intracellular digestion and even divide. But such cells cannot differentiate and their life appears to be restricted to about 20 days or less. From these phenomena it is tempting to suggest that drugs which have an immediate action exercise their effect primarily upon cytoplasmic processes, whereas those with a delayed action have a primary action upon the nucleus. But whilst the first hypothesis may well be correct, more doubt must attach to the second, insofar that there are now reasons for supposing that in addition to the genetical activity of the nucleus we must also consider genetically active particles in the cytoplasm.

As an example of the type of phenomenon which has to be considered may be mentioned the case of lewisite. By using transparent chambers for the study of the skin similar to those designed by CLARK, it has been shown that large or moderate doses of lewisite cause cell death within a few hours. But when a very small dose of lewisite is applied to the skin, there is a transient effect which passes off after about two hours, and may be followed several days later by a much more profound effect which may result in the death of cells after about a week (Fig. 21). It is very probable indeed that the destruction of

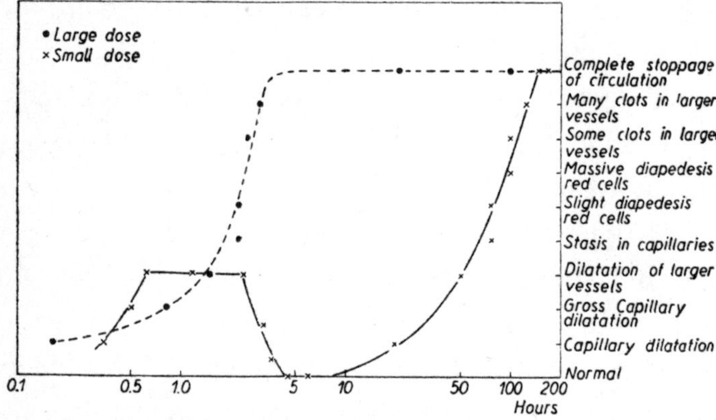

Fig. 21. The response of the skin of the ear of rabbits to large and small doses of lewisite

cells caused by relatively large doses of lewisite, and the transient interference with cellular activity observed soon after slight contamination, can be attributed to cytoplasmic damage. But whether the secondary phenomena developing with slight contamination are due to cytoplasmic or nuclear damage is much more difficult to decide. The secondary phenomena are probably closely related to systemic poisoning by arsenical compounds, so that the solving of this problem is not necessarily of academic interest only.

Possible Modes of Drug Action upon Genes

At least three consequences may follow the action of a drug upon a gene. 1. The normal physiological action of the gene may be reduced or abolished. 2. Mutation of the gene may occur. 3. Reproduction of the gene may be inhibited. If a cell changes its behaviour or nature under the action of a drug it may or may not return to its initial condition when the drug is removed. If the action of the drug is reversible, its action is commonly said to be physiological, whether the action is upon a gene or not. If the action is irreversible it is thought to involve a mutation, although it is usually only possible to prove this when sex cells are involved.

It is of much interest that the same or similar substances may often be simultaneously 1. morphogenetic evocators, 2. carcinogenic and 3. hormones with a physiological function. For example, members of the oestrogen series have all three of these activities. At present one may well be inclined, on theoretical grounds, to suggest that 1. and 2. are in fact similar processes involving the mutation of genes.

The situation is complicated by the fact that, although most of the genetic phenomena with which we are accustomed to deal are mediated by genes attached to chromosomes and obeying the Mendelian laws, evidence is steadily accumulating to show that some transmissbile characteristics are carried, not necessarily by nuclear

Cell Physiology 9*

genes, or not exclusively by nuclear genes, but also by genetically active bodies in the cytoplasm, or plasma-genes. Thus, even in the cases where it is suspected tht mutation has occurred in sex cells, it is not always possible to test for this by the normal procedures which can be used for studying mutation of nuclear genes.

HADDOW has suggested that mutations may rather commonly take place under the action of growth inhibitors and that this mutation may take the form of permitting the escape of a cell from the action of an inhibitor. In the normal animal growth in most organs in the adult is restrained to just that degree which is necessary to permit replacement of cells which have died. The process by which this control is established is very far from fully understood. But it is known that there are some substances present in animal tissues which promote cell growth and cell division, and others which inhibit this process. It seems likely that the growth-promoting substances are sometimes in some sense used up by the cells upon which they act, so that in the case of the substances which promote cell division there tends to be an equilibrium established between the concentration of the substance and the number of cells acting upon it. The action of inhibitors upon this process is likely to be in the direction of shifting the equilibrium position so that fewer cells are in equilibrium with a given concentration of growth promoter. There are a great many possible ways in which the inhibitory substances might produce such an effect, amongst the more entertaining of which is the possibility that the inhibitors may in fact be substances causing a more rapid turn-over of the growth-promoting substance in the cells. It is a characteristic of this balance of control that it is adjusted so that normally mutations permitting escape from this control are infrequent. But when additional foreign growth inhibitors are present, mutation appears to be more frequent, and may involve not only escape from the foreign growth inhibitor but also from the normal endogenous growth inhibitors. Cells which suffer such mutations may give rise to tumours. There is a good deal of evidence available now suggesting that mutations, whether they be of nuclear genes, or of plasma genes, may commonly take place under the action of drugs, such as penicillin, the sulphonamides, arsenicals, etc. and thus give rise to strains of cells (usually of micro-organisms) which are resistant to the drug concerned.

If we are seriously to adopt the point of view which has just been suggested, namely, that mutation may be a fairly common event under appropriate conditions, we must reconsider our attitude towards the stability of genes. It has usually been supposed that genes are remarkably stable bodies. This point of view has arisen because the occurrence of mutations is normally a very infrequent process. However, when we consider the probable chemical composition of genes, i.e. a combination of deoxypentose nucleic acid and protein, there appears to be no intrinsic probability that genes are by nature very stable. On the contrary both the proteins and the deoxypentose nucleic acids are very unstable towards suitable reagents. Early attempts to produce mutations by the action of chemicals were a failure, and these failures tended to reinforce the view that genes were remarkably stable. But consideration of the chemicals which were studied shows that practically without exception they were substances which either would never get into a cell without first killing it by destroying the permeability of the plasma membrane, or else were substances which were certain to undergo almost instantaneous reaction with the cytoplasm after entering the cell. I.e. the substances studied were almost all singularly unlikely to make contact with the nuclear genes. More recent experiments, starting with those of ROBSON, AUERBACH and KOLLER on mustard gas, have shown that chemical substances which have physical properties which will both enable them to penetrate into the cell nucleus and to react with the components of genes are remarkably effective in producing mutations, thus confirming the evidence obtained by the study of radiations, which also have the property of being able to penetrate into the nucleus and secure a reaction with gene components. We may thus conclude that the stability normally exhibited by genes is not an intrinsic refractoriness towards change, but is attributable to the genes being present in a very stable environment. Towards suitable reagents we may expect genes to be very unstable, and it is possible that the morphogenetic evocators and the more potent carcinogens are such reagents.

If the hypothesis just formulated should prove to be correct, we should have a profitable line of approach to a number of biological problems. As a particularly interesting example we may take the work of BERENBLUM on the induction of cancer by croton oil and dimethyl benzanthracene. When croton oil is applied alone to the skin of a mouse hyperplasia rapidly develops, but normally no tumours will develop during the natural life of the animal. If dimethyl benzanthracene alone is applied to the skin of a mouse hyperplasia results and after some period tumours commonly develop. If both croton oil and dimethyl benzanthracene are applied together, tumours appear significantly more rapidly than when dimethyl benzanthracene is applied alone. Even more striking is the fact that if just one application of dimethyl benzanthracene is made, tumours appear after a rather protracted period. But if, after the application of dimethyl benzanthracene, croton oil is applied, tumours develop much more rapidly and there is a strong tendency for the appearance of the tumour to occur at a standard time after the application of croton oil, rather than after the application of dimethyl benzanthracene. At present the only possible interpretation of these results is that dimethyl benzanthracene rather readily causes an irreversible change in the cells to which it is applied, which remains latent until the damage is revealed by an irritant such as croton oil. We could readily understand these results if the irreversible damage caused by the dimethyl benzanthracene is a mutation.

The Relationship between Hormones and Evocators

At present it is supposed that hormones are usually in some way connected with the activities of enzymes, and exercise their physiological effect through this connection. The hormones may be prosthetic groups, coenzymes or inhibitors etc. of the enzymes concerned. We also suspect that genes consist of a specific array of enzymes. This being so, we are inclined to suspect that hormones and genes may have some fairly direct relationships. A possible source of the relationship between the evocator action and the physiological hormone action of a given substance can be traced if we consider the circumstances of gene reproduction.

One of the most striking characteristics of genes is their ability to reproduce themselves. We do not know a great deal about the conditions which are necessary for the control of reproduction of genes. But two possible extreme cases can be distinguished. Some genes are perhaps in a cell quite intact at all times, and reproduce themselves in the general environment provided by the cell. Other genes may need additional specific substances present, which may be called primer substances, before they can reproduce. Now if a gene combines with a hormone it is no longer exactly the same body as it was before that combination took place, and it is obvious that it may be unable to reproduce itself during the period in which it is changed by combination with the hormone. Another possibility is that if it does reproduce, it will not reproduce exactly as it was before combination with the hormone, but as a new body: i.e. it may reproduce as a mutant of the original gene. With these points is mind we can see that the nature of the action of a hormone upon a gene may depend upon the fraction of time with which gene is combined with hormone. If the concentration of hormone is less than a roughly defined concentration, the hormone will affect only the physiological activity of the gene. But if the concentration of the hormone exceeds this rough level, the gene may be unable to reproduce, or may reproduce a mutant gene. The possibility of a hormone producing a mutation of a gene by combination with it is itself very interesting. When we consider the case of a primer substance also being required for gene reproduction a number of related alternative mechanisms for the production of mutant genes appear.

Similarly, if the action of the hormone is directly or indirectly to prevent gene reproduction, then at least two possibilities emerge. One is that the gene or primer may be destroyed in the cell during the period in which it is unable to reproduce. In this case it is treated virtually as a foreign body. Alternatively, cell division may proceed until a cell is produced which lacks the gene or primer which cannot reproduce itself. Both possibilities result in the appearance of a cell which is lacking the gene upon which the hormone acts, i.e. deletion of the gene has occurred.

Mutation or deletion of a gene need not become obvious immediately. For example, let us consider the case of a chromosome gene, which exercises its physiological effect by producing a product which is itself competent to reproduce, e.g. the product is a plasmagene. Then, destruction of the chromosome gene may only appear if, 1. later circumstances also prevent self-reproduction of the plasmagene, or 2. later circumstances result in mutation of the plasmagene. In fact damage to a gene caused by one circumstance may become apparent only very much later, as the result of some other circumstance quite unrelated to the first, and perhaps many years may elapse before the second circumstance occurs. It may be that it is phenomena such as these which account for the appearance of a tumour after the elapse of many years from the time of exposure to a carcinogenic agent. And phenomena of this type may be involved in results such as have been described above with croton oil and dimethyl benzanthracene. I.e. croton oil may perhaps reveal the damage caused to a gene by dimethyl benzanthracene, through an action upon a plasmagene which is the normal product of the gene upon which dimethyl benzanthracene has its effect.

There are many other problems which might well be investigated from points of view analogous to that which we have been discussing. For example, some types of leukemia respond well to treatment with urethan for a period, but after that period become resistant to urethan. Is some phenomenon of mutation or deletion involved here? Then we might, for example, ask whether a sulphonamide-induced agranulocytosis is an instance of gene deletion? But space does not permit of going into further details in such matters.

REFERENCES

AUERBACH, C. and ROBSON, J. M., 1944: *Nature*, 154, 81.

BERENBLUM, I. and SHUBIK, P., 1947: *Brit. J. Cancer*, 1, 379.

BLAKESTEE, A. F. and AVERY, A. G., 1937: *J. Hered.*, 28, 393.

CLARK, A. J., 1933: *Mode of Action of Drugs on Cells* (Arnold, London).

CLARK, A. J., 1937: *General Pharmacology* (Handbuch der Exp. Pharm., IV).

DANIELLI, J. F., 1940: In *Cytology and Cell Phys.* Edit. Bourne (Clarendon, Press, Oxford).

DANIELLI, J. F. and CATCHESIDE, D. G., 1945: *Nature*, 156, 294.

DARLINGTON, C. D. and KOLLER, P. C., 1947: *Heredity*, 1, 187.

DUSTIN, A. P., 1925: *C. r. Soc. Biol.*, 93, 465.

DUSTIN, P., 1947: *Nature*, 159, 794.

EYNNY, H., JOHNSON, F. H. and GENSLER, R. L., 1946: *J. Phys. Chem.*, 50, 453.

GALE, E. F., 1949: *Symp. Soc. Exp. Biol.* III, 233.

GOMORI, G., 1939: *Proc. Soc. Exp. Biol. and Med.*, 42, 23.

HERRIOT, R. M., 1948: *J. Gen. Physiol.*, 32, 221.

HUGHES, A. and FELL, H. B., 1949: *Quart. J. Micros. Sci.*, 90, 37.

JOHNSON, F. H., BROWN, D. E. and MARSLAND, D., 1942: *J. Cell. Comp. Physiol.*, 20, 247, 269.

KARUSH, F. and SIEGEL, B. M., 1948: *Science*, 108, 107.

KOLLER, P. C., 1947: *Symp. Soc. Exp. Biol.*, I, 270 (Cambridge Univ. Press London).

KOLLER, P. C., 1947: *Brit. J. Cancer*, 1, 38.

NEEDHAM, J., 1942: *Biochemistry and Morphogenesis* (Cambridge Univ. Press, London).

OSTERGREN, G. and LEVAN, A., 1943: *Hereditas*, 29, 496.

OSTERGREN, G., 1944: *Hereditas*, 30, 429.

REVELL, S. and LOVELESS, A., 1949: *Nature*, 164, 938.

ROTHEN, A. 1946: *J. Biol. Chem.*, 163, 345.

ROTHEN, A. 1947: *J. Biol. Chem.*, 167, 299.

TAKAMATSU, H., 1939: *Trans. Soc. Path. Japan*, 29, 492.

WOODS, D. D. and NIMMO-SMITH, R. H., 1949: *Symp. Soc. Exp. Biol.* III, 177.

Author Index

Abramson, H. A., 45
Adam, N. K., 24, 45
Adrian, E. D., 90
Alexander, A. E., 42, 45
Auerbach, C., 93, 96, 144, 149
Avery, A. G., 149

Baldwin, E., 96
Bärlund, A., 73, 102, 114
Berenblum, I., 145, 149
Bergman, 76
Bernal, J. D., 13, 24, 45
Blakestee, A. F., 149
Bourne, G., 24
Brachet, J., 4, 24
Brinley, F. J., 100, 114
Brown, D. E., 114, 149

Cameron, 93
Cannan, 92
Catcheside, D. G., 9, 149
Clark, A. J., 24, 73, 74, 96, 114,
    139, 149
Collander, R., 49, 73
Cori, 92
Cullumbine, H., 96

Dale, H., 38, 45
Danielli, J. F., 9, 24, 27, 31, 45,
    114, 149
Darlington, C. D., 24, 149
Davies, J. T., 24, 45
Davson, H., 24, 73, 102, 114
Dixon, M., 89, 91, 92, 93, 96
Dustin, A. P., 149
Dustin, P., 149

Eynny, H., 149

Fankuchen, A., 24, 43, 45
Fell, H. B., 149
Feulgen, 134
Fildes, 136

Gale, E. F., 136, 149
Gensler, R. L., 149
Gilman, A., 96
Gomori, G., 134, 149
Gray, J., 24

Haddow, A., 142
Hawking, 67
Herriot, R. M., 135, 149
Hiller, S., 100, 114
Höber, R., 101, 114
Hughes, A., 149

Jacobs, 102
Johnson, F. H., 113, 114, 149

Karush, F., 149
Keilin, D., 74
King, 67
Koller, P. C., 93, 96, 133, 144,
    149, 150

Levan, A., 150
Lillie, R. S., 73, 101, 114
Loveless, A., 150
Marsland, D., 100, 101, 114, 149
McAnally, M., 73
Meyer, K. H., 105, 106, 113, 114
Mirsky, A. E., 8, 24
Moore, 5

Needham, D. M., 91, 92, 93, 96, 150

Nimmo-Smith, R. H., 150

Östergren, G., 126, 150

Overton, 106, 113

Parpart, 102

Peters, R. A., 69, 73, 74, 96

Phillips, G. S., 96

Phillipson, J., 73

Quastel, J. H., 69, 73, 111, 112, 114

Rapkine, 131

Revell, S., 133, 150

Rideal, E. K., 33, 45

Ris, H., 24

Robson, J. M., 93, 96, 144, 149

Rothen, A., 45, 150

Saunders, 90

Schulman, J. H., 33, 45

Shubik, P., 149

Siegel, B. M., 149

Stedman, E., 8, 24

Stocken, J. R., 69, 73

Takamatsu, H., 134, 150

Thompson, R. H. S., 69, 73

Traube, I., 102, 105, 114

Trim, A. R., 42, 45, 66

Voegtlin, F. R., 69, 73

Webb, D. A., 27, 45

Wilson, E. B., 24

Winterstein, H., 73, 101, 114

Woods, D. D., 136, 150

Work, E., 96

Work, T. S., 96

Activators, 15, 75
Active patches, 19
Acetylcholine, 19
Adrenaline, 34, 138
Agranulocytosis, 149
Alcohols, 98
Amoeboid movement, 5, 86
Anthelminthics, 66
Antibodies, 21, 22
Arsenic acid, 77, 83
Arsenical poisoning, 70
Arsenicals, 143
Arsenite, 88
Arsenoxides, 64, 67, 76, 131
Arterioles, skin, 86
Asters, 13, 123
Azide, 83

Bacteriostasis, 136
B.A.L., 70, 132
—, glucoside, 72
Blood-brain barrier, 64
Brain, 63, 110
— tissue, 112
Butyl alcohol, 102

Calcium, 26, 28, 29, 30
Cancer, 138, 141, 145
Carbon monoxyde, 83
Carriers, 75
Cell division, 5, 39, 86
Cell form, 5
Cellular structure, 12
Centrifugal force, 3
Centrifugation, 4
Centrosome, 124

β-Chloroethylamines, 88
Cholic acid, 64
Choline esterase, 89, 90, 91
Chromomeres, 7, 8
Chromosomes, 7, 8, 15, 44, 123,
    131, 133, 141, 148
Ciliary movement, 86
Cocaine, 100
Coenzymes, 76, 146
—, II, 76
Colchicine, 88, 126
Cortical gels, 13, 123
Croton oil, 145, 148
Curare, 100
Cytochemical studies, 2, 6, 8
Cytochemistry of hepatic cells, x
Cytochrome, 75
— system, 82
Cytolysis, 36

Defence mechanisms, 20
Dehydrogenases, 75, 76, 77, 89
Detoxication, 20
Dichlorophenol, 110
Dielectric constant of cellular
    systems, 17
Differentiation, 44
Diffusion, 46, 53, 54, 63
Dimethyl benzanthracene, 145,
    148
Dithioglycerol, 88
Dithiol, 69

Enzyme(s), 15, 17, 19, 21, 23, 69,
    76, 78, 109, 134, 146
—, action, 44, 48

Enzyme, cellular, 74
— poisons, 60
Equatorial plate, 124
Eserine, 90
Ethyl iodoacetate, 88
Evocators, 141, 146

Fluoride, 76, 83
Fluorophosphonates, 91

Gel, 5
—, cortical, 13, 123
—, protoplasmic, 4
Genes, 8, 11, 16, 19, 23, 123,
    134, 137, 141, 143, 144, 146,
    148
Glucose, 83
Glyceraldehyde, 76, 83
Glycolysis, 82, 86
Granules, 14, 15, 20

Heavy metals, 32
—, oligodynamic effect, 30
Hexokinase, 89, 90, 91, 92, 95
Hexyl resorcinol, 43
High pressure, 113
Hormones, 141, 146, 148
Hydrocyanic acid, 75, 77, 81, 83,
    84, 85, 100, 132
Hydrogen sulphide, 81, 100

Inhibitors, 15, 75, 142, 146
—, competitive, 76
Iodine, 76
Iodoacetamide, 131
Iodoacetate, 76, 83, 85, 88
Ions, 25, 27, 29, 30, 53, 102
Isotopes, 19

Lachrymation, 88
Lachrymators, 89
Leucotaxin, 92

Leukemia, 149
Lewisite, 69, 70, 83, 84, 91, 92,
    139
Lipoid membrane, 61
— molecules, 14
— solubility, 63
Lipoids, 36
Lysis, 36

Magnesium, 100
Malonate, 76
Mastitis in cattle, 64
Membrane, cell, 101
—, lipoid, 61
—, nuclear, 18, 124
—, plasma, 36
— properties, 93
Membranes, 15
—, natural, 46
Metals, heavy, 32
Methylene blue, 75
Micelles, 14, 42
— formation, 39, 40
Mitosis, 4, 13, 93, 123, 126, 133
    136, 137
Monolayer, 26, 32
—, protein, 37, 38
Muscle, 82, 120
— cell, 88
— cells, striated, 6
— contraction, 86
Mustard, 131, 134
Mustard gas, 83, 84, 85, 91, 93, 144
Mutation, 141, 146, 148
Mutations, 74, 142, 143
Myosin, 6
Myotics, 91

Narcotics, 97
Nerve, 120
Nicotinic acid, 76
Nitrogen mustards, 76

Nuclear membrane, 18, 124
Nuclei, 94, 139
Nucleic acids, 2, 6, 14, 15, 16,
    19, 124, 143
Nucleoli, 15, 124
Nucleus, 5, 16, 93, 133

Oestrogens, 37, 38
Oestrone, 138
Oligodynamic action, 32
— effect of heavy metals, 30
Osmic acid, 88
Oxygen, 83

Paraffins, 101
Parthenogenesis, 122
Partition coefficients, 126
— effects, 105
Penicillin, 136, 143
Permeability, 18, 22, 48, 50, 52,
    54, 58, 60, 63, 66, 68, 92, 99,
    101, 144
pH, 11
— buffers, 16
Phagocytosis, 5
Phenol, 79
Phenols, 132
Phosphokinases, 92
Picric acid, 100
Plasmagenes, 142, 148
Plasma membrane, 36
Polarisation, 18
Potassium, 88
Prosthetic groups, 76, 146
Proteases, 92
Proteins, 8, 14, 15, 16, 19, 28, 36,
    74, 113, 143
—, monolayer, 37, 38
—, spindle, 132
Protoplasmic streaming, 5
Prussic acid, see Hydrocyanic
    acid

Purple, visual, 19
Pyocyanin, 75
Pyridoxal, 76
Pyruvic oxidase, 92, 93

Quinones, 133

Radiation sickness, 93
Radiations, 144
Reactions at interfaces, 33
Removers, substrate, 77
Renal secretion, 86
Resistance, 21, 143
Respiration, 82, 110
—, cellular, 86
Rumen, 59

Secretion, 46, 60, 94
Secretory activity, 22
Self-reproduction, 23
Silver, 48
Slime moulds, 5
Soap, 88, 90
Spindle(s), 13, 124, 131, 133
— proteins, 132
Structure, cellular, 12
Substrates, 15
Substrate removers, 77
Sulphonamides, 76, 88, 136, 143,
    149
Surface action, 15
— tension, 102
Surfaces, 99, 103, 104
— of skin cells, 92

Tactoids, 13
Trypanosomes, 67, 68
Tryptophan, 138

Urethan, 83, 88, 110, 131, 132,
    149

Vacuoles, 20
Vesicant substances, 95
Vesicants, 90, 91
Vesication, 70, 120
Viruses, 23, 134, 137

Visual purple, 19
Vitamin B_1, 76, 83
Vitamin B_2, 76

Xanthopterin, 138